GEOTARGETED
ALERTS AND WARNINGS

REPORT OF A WORKSHOP ON CURRENT
KNOWLEDGE AND RESEARCH GAPS

Committee on Geotargeted Disaster Alerts and Warnings:
A Workshop on Current Knowledge and Research Gaps

Computer Science and Telecommunications Board

Division on Engineering and Physical Sciences

NATIONAL RESEARCH COUNCIL
OF THE NATIONAL ACADEMIES

THE NATIONAL ACADEMIES PRESS
Washington, D.C.
www.nap.edu

THE NATIONAL ACADEMIES PRESS   500 Fifth Street, NW   Washington, DC 20001

NOTICE: The project that is the subject of this report was approved by the Governing Board of the National Research Council, whose members are drawn from the councils of the National Academy of Sciences, the National Academy of Engineering, and the Institute of Medicine.

Support for this project was provided by the Department of Homeland Security under contract number HSHQDC-11-D-00009, task order number HSHQDC-12-J-00189. Any opinions, findings, or conclusions expressed in this publication are those of the authors and do not necessarily reflect the views of the organizations that provided support for the project.

International Standard Book Number-13   978-0-309-28985-6
International Standard Book Number-10:   0-309-28985-8

This report is available from:

Computer Science and Telecommunications Board
National Research Council
500 Fifth Street, NW
Washington, DC 20001

Additional copies of this report are available from the National Academies Press, 500 Fifth Street, NW, Keck 360, Washington, DC 20001; (800) 624-6242 or (202) 334-3313; http://www.nap.edu.

Copyright 2013 by the National Academy of Sciences. All rights reserved.

Printed in the United States of America

THE NATIONAL ACADEMIES
Advisers to the Nation on Science, Engineering, and Medicine

The **National Academy of Sciences** is a private, nonprofit, self-perpetuating society of distinguished scholars engaged in scientific and engineering research, dedicated to the furtherance of science and technology and to their use for the general welfare. Upon the authority of the charter granted to it by the Congress in 1863, the Academy has a mandate that requires it to advise the federal government on scientific and technical matters. Dr. Ralph J. Cicerone is president of the National Academy of Sciences.

The **National Academy of Engineering** was established in 1964, under the charter of the National Academy of Sciences, as a parallel organization of outstanding engineers. It is autonomous in its administration and in the selection of its members, sharing with the National Academy of Sciences the responsibility for advising the federal government. The National Academy of Engineering also sponsors engineering programs aimed at meeting national needs, encourages education and research, and recognizes the superior achievements of engineers. Dr. C. D. Mote, Jr., is president of the National Academy of Engineering.

The **Institute of Medicine** was established in 1970 by the National Academy of Sciences to secure the services of eminent members of appropriate professions in the examination of policy matters pertaining to the health of the public. The Institute acts under the responsibility given to the National Academy of Sciences by its congressional charter to be an adviser to the federal government and, upon its own initiative, to identify issues of medical care, research, and education. Dr. Harvey V. Fineberg is president of the Institute of Medicine.

The **National Research Council** was organized by the National Academy of Sciences in 1916 to associate the broad community of science and technology with the Academy's purposes of furthering knowledge and advising the federal government. Functioning in accordance with general policies determined by the Academy, the Council has become the principal operating agency of both the National Academy of Sciences and the National Academy of Engineering in providing services to the government, the public, and the scientific and engineering communities. The Council is administered jointly by both Academies and the Institute of Medicine. Dr. Ralph J. Cicerone and Dr. C. D. Mote, Jr., are chair and vice chair, respectively, of the National Research Council.

www.national-academies.org

**OTHER RELATED REPORTS OF THE COMPUTER
SCIENCE AND TELECOMMUNICATIONS BOARD**

Public Response to Alerts and Warnings Using Social Media: Summary of a Workshop on Current Knowledge and Research Gaps (2013)

Public Response to Alerts and Warnings on Mobile Devices: Summary of a Workshop on Current Knowledge and Research Gaps (2011)

Improving Disaster Management: The Role of IT in Mitigation, Preparedness, Response, and Recovery (2007)

The Internet Under Crisis Conditions: Learning from September 11 (2003)

Limited copies of CSTB reports are available free of charge from
Computer Science and Telecommunications Board
National Research Council
The Keck Center of the National Academies
500 Fifth Street, NW, Washington, DC 20001
(202) 334-2605/cstb@nas.edu
www.cstb.org

COMMITTEE ON GEOTARGETED ALERTS AND WARNINGS: A WORKSHOP ON CURRENT KNOWLEDGE AND RESEARCH GAPS

ELLIS M. STANLEY, SR., Independent Consultant, Atlanta, Georgia, *Chair*
ART BOTTERELL, Carnegie Mellon University, Silicon Valley
K. MANI CHANDY, California Institute of Technology
DENNIS S. MILETI, University of Colorado, Boulder
HELENA MITCHELL, Georgia Institute of Technology
RAMESH R. RAO, University of California, San Diego
SHASHI SHEKHAR, University of Minnesota, Minneapolis
MING-HSIANG TSOU, San Diego State University

Staff

VIRGINIA BACON TALATI, Program Officer
SHENAE BRADLEY, Senior Program Assistant
LINDA WALKER, Senior Program Assistant

JON EISENBERG, Director, Computer Science and Telecommunications Board

COMPUTER SCIENCE AND TELECOMMUNICATIONS BOARD

ROBERT F. SPROULL, Oracle (retired), *Chair*
JACK L. GOLDSMITH III, Harvard Law School
SEYMOUR E. GOODMAN, Georgia Institute of Technology
ROBERT KRAUT, Carnegie Mellon University
SUSAN LANDAU, Radcliffe Institute for Advanced Study
PETER LEE, Microsoft Corporation
DAVID E. LIDDLE, US Venture Partners
JOHN STANKOVIC, University of Virginia
JOHN A. SWAINSON, Dell, Inc.
PETER SZOLOVITS, Massachusetts Institute of Technology
ERNEST J. WILSON, University of Southern California
KATHERINE YELICK, University of California, Berkeley

Staff

JON EISENBERG, Director
VIRGINIA BACON TALATI, Program Officer
SHENAE BRADLEY, Senior Program Assistant
RENEE HAWKINS, Financial and Administrative Manager
HERBERT S. LIN, Chief Scientist, CSTB
LYNETTE I. MILLETT, Associate Director, CSTB
ERIC WHITAKER, Senior Program Assistant

For more information on CSTB,
see its website at http://www.cstb.org, write to CSTB,
National Research Council, 500 Fifth Street, NW, Washington, DC 20001,
call (202) 334-2605, or e-mail the CSTB at cstb@nas.edu.

Preface

Following earlier workshops organized by separate National Research Council (NRC) committees that explored the public response to alerts and warnings delivered to mobile devices[1] and alerts and warnings delivered using social media,[2] the Department of Homeland Security's Science and Technology Directorate asked the Computer Science and Telecommunications Board of the NRC to convene a workshop to examine more precise geotargeting of alerts and warnings. This report presents a summary of a February 21-22, 2013, workshop organized by the NRC's Committee on Geotargeted Disaster Alerts and Warnings: Current Knowledge and Research Gaps. The workshop brought together social science researchers, technologists, emergency management professionals, and other experts to explore (1) what is known about how the public responds to geotargeted alerts and warnings; (2) technologies and techniques for enhancing the geotargeting of alerts and warnings; and (3) open research questions about how to effectively use geotargeted alerts and warnings and technology gaps. The complete statement of task for the workshop is provided in Box P.1, and the workshop agenda is provided in Appendix A.

---

[1] National Research Council, *Public Response to Alerts and Warnings on Mobile Devices: Summary of a Workshop on Current Knowledge and Research Gaps,* The National Academies Press, Washington, D.C., 2011.

[2] National Research Council, *Public Response to Alerts and Warnings Using Social Media: Report of a Workshop on Current Knowledge and Research Gaps,* The National Academies Press, Washington, D.C., 2013.

> **BOX P.1**
> **Statement of Task**
>
> An ad hoc steering committee will plan and conduct a public workshop that will consider the potential for more precise geographical targeting to improve the effectiveness of disaster alerts and warnings; examine the opportunities presented by current and emerging technologies to create, deliver, and display alerts and warnings with greater geographical precision; consider the circumstances where more granular targeting would be useful; and examine the potential roles of federal, state, and local agencies and private sector information and communications providers in delivering more targeted alerts. The committee will organize the workshop to include a mix of individual presentations, panels, breakout discussions, and question-and-answer sessions to develop an understanding of the relevant research communities, research already completed, ongoing research, and future research needs. Key stakeholders would be identified and invited to participate. The committee will develop the workshop agenda, select and invite speakers and discussants, and moderate the discussions. An unedited (verbatim) transcript of the event would be prepared. A report summarizing the committee's assessment of what transpired at the workshop would also be prepared.

This report summarizes presentations made by invited speakers and other remarks by workshop participants. In keeping with the workshop's purpose of exploring an emerging topic, this summary does not contain findings or recommendations. Nor, in keeping with NRC guidelines for workshop reports, does it necessarily reflect consensus views of the workshop participants or the organizing committee. The summary draws on the prepared remarks of workshop presenters, comments made by workshop participants, and the ensuing discussion.

The first two chapters of this report summarize presentations and discussions on the value of geotargeted alerts and warnings (Chapter 1) and technologies and tools for geotargeted alerts and warnings (Chapter 2). Chapter 3 summarizes the research questions—reflecting gaps in our present understanding—that were identified by workshop participants during the course of the workshop. Appendix A provides the workshop agenda, and speaker biosketches are provided in Appendix B. Appendix C provides biosketches of the committee members.

Ellis Stanley, *Chair*
Committee on Geotargeted Disaster Alerts and Warnings:
Current Knowledge and Research Gaps

Acknowledgment of Reviewers

This report has been reviewed in draft form by individuals chosen for their diverse perspectives and technical expertise, in accordance with procedures approved by the National Research Council's Report Review Committee. The purpose of this independent review is to provide candid and critical comments that will assist the institution in making its published report as sound as possible and to ensure that the report meets institutional standards for objectivity, evidence, and responsiveness to the study charge. The review comments and draft manuscript remain confidential to protect the integrity of the deliberative process. We wish to thank the following individuals for their review of this report:

Hisham Kassab, MobiLaps, LLC,
Robert Kraut, Carnegie Mellon University,
Leslie Luke, San Diego County Office of Emergency Services,
Patrick P. Meier, Qatar Foundation Computer Research Institute,
Richard Muth, Baltimore County School District,
Brenda Phillips, Oklahoma State University, and
Robert Sproull, Oracle (retired).

Although the reviewers listed above have provided many constructive comments and suggestions, they were not asked to endorse the conclusions or recommendations, nor did they see the final draft of the report before its release. The review of this report was overseen by Joseph Traub,

Columbia University. Appointed by the National Research Council, he was responsible for making certain that an independent examination of this report was carried out in accordance with institutional procedures and that all review comments were carefully considered. Responsibility for the final content of this report rests entirely with the authoring committee and the institution.

Contents

1   **THE ROLE OF GEOTARGETED ALERTS AND WARNINGS IN DISASTER RESPONSE**   1
Current and Future Vision for the Integrated Public Alert and Warning System, 3
Some Current Knowledge and Research on Geotargeted Alerts and Warnings, 8
Hazard Type and Geotargeting, 15
Data Security and Privacy Challenges, 19

2   **TECHNOLOGIES AND TOOLS FOR GEOTARGETED ALERTS AND WARNINGS**   24
Continuing Opportunities for Using Traditional Technologies for Geotargeted Alerts and Lessons for the Use of New Technologies, 25
Technologies for Geotargeting Alerts Over the Internet, 29
Mobile Device Location Determination Capabilities, 31
Current and Future Technologies for Geotargeting Alerts to Mobile Devices, 32

3   **RESEARCH NEEDS AND IMPLEMENTATION CHALLENGES**   35
Facilitating and Improving Public Response, 35
Value of Geotargeted Information, 36
Developing and Deploying Technology, 37
Respecting Privacy and Meeting Security Needs, 38
Facilitating and Encouraging Use by Practitioners, 38

APPENDIXES

A Workshop Agenda   41
B Biosketches of Workshop Speakers   46
C Biosketches of Committee Members   60

1

The Role of Geotargeted Alerts and Warnings in Disaster Response

This report summarizes a February 2013 workshop organized by the National Research Council's Committee on Geotargeted Disaster Alerts and Warnings: Current Knowledge and Research Gaps. In the context of alerts and warnings,[1] geographical targeting, or geotargeting, refers to the effort to transmit alerts only to those recipients physically located in a geographic area affected by an event and/or are at risk.

Geotargeting is a two-step process. The first is the geo-definition of the targeted area is (e.g., a county or a geographic polygon). The second is the geo-delivery of the message to recipients within the targeted area. Both steps are susceptible to imprecision. In the case of geo-definition, some alerting systems only allow the definition of an area down to the county level, even if the actual affected area is only a section within it. In the case of geo-delivery, receipt of wireless transmissions cannot be limited precisely to a specific region. Any imprecision in one or both steps results in imprecision in the overall geotargeting.

At a 2011 CSTB workshop, it was observed that "[l]ocalization of

---

[1] An *alert* notifies the recipient that something significant has happened or may happen, and a *warning*, which typically follows an alert, provides more detailed information describing the event and indicates what protective action should be taken by the recipient. The distinction between alerts and warnings is not always clear-cut because a warning can also serve as an alert, and an alert may include some information about protective measures. Technology has further eroded the distinction. For example, sirens have evolved to provide both a siren sound (the alert) and a spoken message (which, depending on how much detail it contains, might be considered a warning).

. . . messages by county or equivalent jurisdiction might be too coarse-grained, especially in the case of large counties and highly localized events."[2] Past research has shown that specific and clear information, including which locations are and are not at risk, increases the likelihood that people take protective action. The less precise the geotargeting, the more likely the recipient will ignore the alert, or choose to opt out of the alerting system, because they are not sure whether the message applies to them. When alerts and warnings are delivered to broader populations than those actually affected by an event, more people than are actually at risk may be sent messages to take action. This chapter explores discussions held by panel members and attendees of the February 2013 workshop on public response to geotargeted alerts and warnings.

There are several systems that provide geotargeted alerts today, with various degrees of precision. Some, such as the so-called "reverse-911" systems, can dial groups of landline telephone subscribers and can achieve a high degree of precision because they are capable of calling subscribers located within a specific polygon. On the other hand, wireless-based systems, such as the Emergency Alert System (EAS), are inherently less precise because of the wireless fencing issue. In the case of the national Wireless Emergency Alerts (WEA) system, alerts are transmitted to cellular phones using cellular broadcast technology.[3] Only cellular towers mapped to the geo-defined region broadcast the message. However, the geotargeting precision of WEA in its initial rollout was additionally affected by a design decision to limit geo-definition to the county level. Another system is the National Oceanic and Atmospheric Administration (NOAA) weather radio, which uses dedicated radio frequencies and special-purpose receivers. It delivers weather and other hazard alerts and allows users to limit alarms to only those alerts designated for their location by specifying regions that are largely aligned with counties or portions of counties. Other services allow recipients to subscribe to alerts for geographic areas of interest to them, regardless of their actual physical location. Generally, such systems are not considered to possess true geotargeting capability.

Better localization might be provided by refinements to existing alerting systems or the use of new technologies. For example, some tighter localization is possible with the current WEA technology, but overlaps in coverage of individual cellular towers limit the precision that is possible. Additionally, if alert messages include information about the target

---

[2] National Research Council, *Public Response to Alerts and Warnings on Mobile Devices: Summary of a Workshop on Current Knowledge and Research Gaps*, The National Academies Press, Washington, D.C., 2011.

[3] WEA was formerly known as the Commercial Mobile Alerting System.

region, then software running on the receiving mobile device may be able to filter the alerts based on the device's location. Existing technologies allow mobile phone positions to be determined with sufficient accuracy for most applications, at least outdoors, and emerging technologies promise significant improvements to indoor location measurement. Much of the second day of the workshop was spent exploring changes to existing technologies or new technologies for increasing the precision of geotargeted alerts; these discussions are explored in Chapter 2.

CURRENT AND FUTURE VISION FOR THE INTEGRATED PUBLIC ALERT AND WARNING SYSTEM

The Federal Emergency Management Agency (FEMA) oversees development and operation of the Integrated Public Alert and Warnings System (IPAWS), a national system for delivering emergency alerts and warnings to the public. One component, WEA, which is in the process of being rolled out, delivers alerts to cell phones using a special broadcast channel in cellular telephone systems. The message and associated metadata are formatted according to the Common Alerting Protocol (CAP) standard. Presentations by Denis Gusty from the Science and Technology Directorate of the Department of Homeland Security (DHS S&T), Wade Witmer from FEMA, and Mike Gerber from the National Weather Service (NWS) described the current state of IPAWS and WEA and outlined future plans for both.

The Federal Communications Commission, in cooperation with cellular telephone carriers, established a cell phone alert program, originally known as the Commercial Mobile Alert System. DHS S&T and FEMA have worked with the cell carriers on deployment and testing of the program, which is now known as WEA.[4] A voluntary program for the cellular carriers, it specifies that participating carriers target government-originated emergency alert messages to the cell towers that lie within the counties affected by an alert. Messages are limited to 90 characters, which limits WEA to providing brief alerts to inform recipients that an event is occurring and that they need to seek additional information. It provides three types of messages: alerts issued by the president, alerts involving

---

[4] WEA was established in response to the Warning, Alert, and Response Network Act (P.L. 109-347) passed by Congress in 2006. The Federal Communications Commission proposed and adopted the network structure, operational procedures, and technical requirements for WEA in 2007 and 2008, in cooperation with commercial wireless providers. Subsequently, the Department of Homeland Security Science and Technology Directorate and the Federal Emergency Management Agency have worked with the wireless carriers on deployment and testing of the system. The present design supports only 140-character messages and messages that fit into a single cell broadcast.

imminent threats to life and safety, and AMBER alerts (law enforcement bulletins in child-abduction cases). Mobile device users can opt out of all but the presidential alerts. However, some devices, such as those running the Android operating system, divide the imminent threats into two categories, extreme (e.g., tornados, extreme wind, or tsunami) and severe (e.g., flash flood, dust storm, or blizzard) and permit users to opt out of severe alerts but still receive extreme alerts.

WEA is designed to complement, not replace, other existing dissemination methods/channels supported by the IPAWS architecture (Figure 1.1). The central gateway, the IPAWS Open Platform for Emergency Networks, which is managed by FEMA, provides access to federal, state, local, territorial, and tribal authorities. Emergency managers are issued digital certificates to authenticate their access to the system and use commercial software to prepare and transmit messages.

The CAP standard supports the use of FIPS codes,[5] which desig-

---

[5] FIPS localization is based on standardized codes for county and equivalent geographical entities previously defined in the now-withdrawn Federal Information Processing Standards (FIPS) 6-4 standard and now defined by the American National Standards Institute INCITS 31:200x standard.

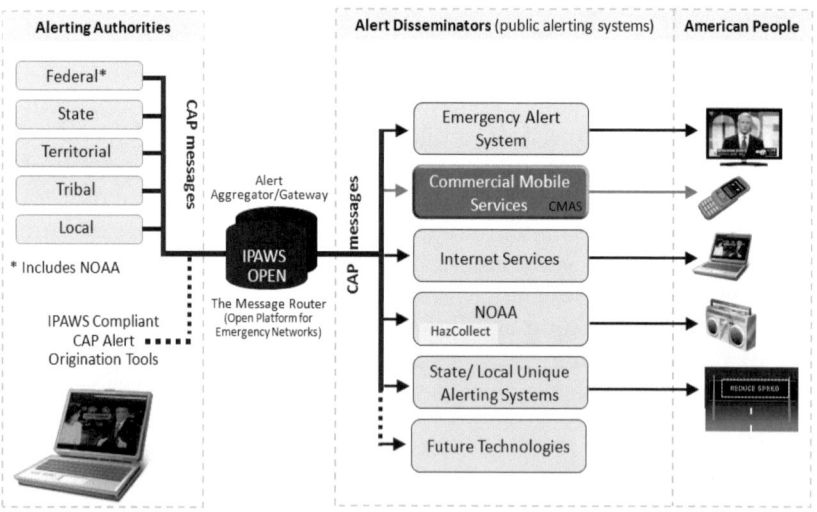

FIGURE 1.1 Structure of the Integrated Public Alerts and Warnings System. Source: D. Gusty, "Overview of the Commercial Mobile Alert System (CMAS) Research and Development Effort," presentation at the Workshop on Geotargeted Alerts and Warnings, Washington, D.C., February 2013.

nate counties and subdivisions within a county, to define the targeted area. (Another standard, specific area message encoding, is still used by weather radios and is discussed in Chapter 2.) CAP also supports the use of a list of polygon vertex coordinates to specify the boundaries of a targeted area. The NWS has for a number of years defined smaller polygons for many of their severe weather alerts, and local agencies have found it beneficial to target alerts to areas much smaller than a county. Carriers have also developed approaches to target smaller areas than required by WEA, although neither perfectly targets the specified polygon (Figure 1.2). Today, each of the three major carriers uses one of these approaches to target only cellular towers in the appropriate portion of a county when subcounty-level alerts are issued.

Current Use of the Integrated Public Alerts and Warning System and the Wireless Emergency Alerts

Today, over 150 entities—including approximately 75 counties, 25 states, and the Commonwealth of Puerto Rico—have access to the IPAWS-OPEN gateway, which allows them to transmit messages to cellular phones (using WEA), radio and television (using the Emergency Alert System), and NOAA weather radios. In addition, the NWS can send weather alerts, and the National Center for Missing and Exploited Children can send AMBER alerts. FEMA receives about four new applications a week for access to IPAWS-OPEN, and the system is projected to grow to include as many as 10,000 entities.

In 2010, the NWS was the first to use the system. Initially, messages were sent only to weather radios, but are sent now to other alerting systems, including WEA. For WEA, the NWS created template text for nine different types of warnings (a small subset of the templates for EAS). These templates are based on those designed for use in EAS but edited to adhere to the CAP standard and limited message length of WEA (Box 1.1). The NWS is currently exploring ways to give forecasters additional capabilities to customize messages and add geographic information.

Current Research Initiatives

DHS's Commercial Mobile Alert Service (CMAS), Research, Development, Testing and Evaluation Program, established as a result of the Warning, Alert, and Response Network Act,[6] is responsible for coordinating WEA testing, developing recommendations and guidance for its use,

---

[6] Title VI of P.L. 109-347.

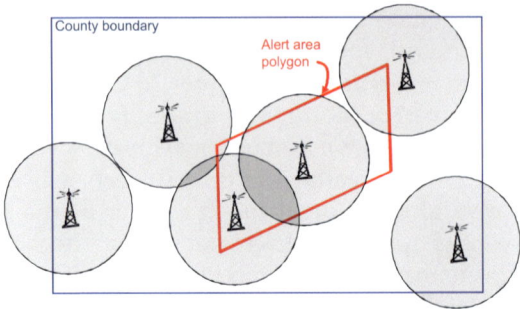

(a) Regulatory requirements require that a carrier targets an entire county area.

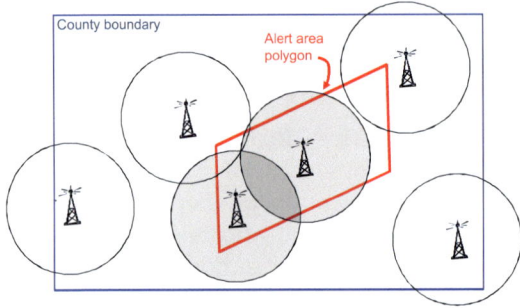

(b) An alert message is transmitted by each cellular tower located within the polygon.

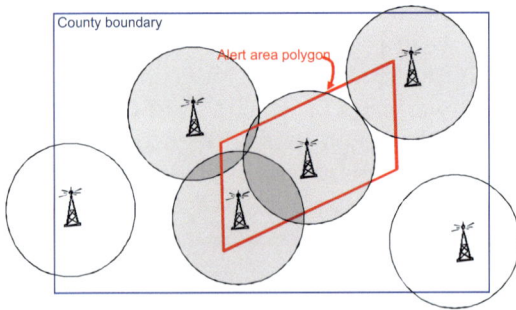

(c) An alert message is transmitted by each cellular tower that provides coverage in the polygon.

FIGURE 1.2 Options for mapping alert polygon regions to cellular towers. The circles surrounding each tower represent that tower's transmission reach; the shading indicates that the tower is sending an alert. SOURCE: W. Witmer, FEMA, "Integrated Public Alerts and Warning System," presentation at the Workshop on Geotargeted Alerts and Warnings, Washington, D.C., February 2013.

> **BOX 1.1**
> **WEA Message Templates Originated by
> the National Weather Service**
>
> | | |
> |---|---|
> | Tsunami warning | Tsunami danger on the coast. Go to high ground or move inland. Check local media.—NWS |
> | Tornado warning | Tornado Warning in this area til hh:mm tzT. Take shelter now. Check local media.—NWS |
> | Extreme wind warning | Extreme Wind Warning this area til hh:mm tzT ddd. Take shelter.—NWS |
> | Flash flood warning | Flash Flood Warning this area til hh:mm tzT. Avoid flooded areas. Check local media.—NWS |
> | Hurricane warning | Hurricane Warning this area til hh:mm tzT ddd. Check local media and authorities.—NWS |
> | Typhoon warning | Typhoon Warning this area til hh:mm tzT ddd. Check local media and authorities.—NWS |
> | Blizzard warning | Blizzard Warning this area til hh:mm tzT ddd. Prepare. Avoid Travel. Check media.—NWS |
> | Ice storm warning | Ice Storm Warning this area til hh:mm tzT ddd. Prepare. Avoid Travel. Check media.—NWS |
> | Dust storm warning | Dust Storm Warning in this area til hh:mm tzT ddd. Avoid travel. Check local media.—NWS |
>
> NOTE: tzT = timezone; ddd = three letter abbreviation for day of the week.
> SOURCE: M. Gerber, National Weather Service, presentation at the Workshop on Geotargeted Alerts and Warnings, Washington, D.C., February 2013.

and funding research and development to enhance alerting systems. The act specifies research in the following specific areas:

- Improving geotargeting of mobile alerts and warnings, including geotargeting granularity, geotargeting in border areas, public response consideration to geotargeting, and alternative technologies to improve current and future geotargeting capabilities; and
- Improving public response to mobile alerts and strategies, including reaching special and diverse populations, information diffusion, and

public response to WEA initiation, message content, message frequency, follow-up, and source.

DHS is currently funding research by the University of Maryland's National Consortium for the Study of Terrorism and Response to Terrorism on comprehensive testing of imminent threat messages for mobile devices and public response and social media (this work is discussed in the section, Current Research on Message Length, Geographical Information, and Geotargeted Alerts) and at the University of Southern Mississippi on geotargeting and other communication approaches for at-risk communities. A broad agency announcement for additional research and development has been issued, and additional awards are expected in 2013.

SOME CURRENT KNOWLEDGE AND RESEARCH ON GEOTARGETED ALERTS AND WARNINGS

In a panel exploring some of the current research on geotargeted alerts and the public response, Timothy Sellnow, University of Kentucky, discussed the role of geotargeted information in effectively communicating risk; Michele Wood, California State University, Fullerton, examined the various ways that geotargeted information can be communicated; Brooke Liu, University of Maryland, discussed hazard types and protective actions as they are related to geotargeted information; and Ken Rudnicki, City of Fairfax, Virginia, discussed tools used by emergency managers to geotarget alerts and warnings.

Wood summarized the results of decades of research on what types of alert and warning message content will motivate appropriate and timely action by the public (Box 1.2). Wood also discussed sense-making, a human process that occurs in the time between the receipt of an alert or warning and the time the recipient takes protective action. Message recipients will try to make sense of the message, confirm the message by seeking additional information, personalize the message to their own circumstances, and only then decide what action they should take. A key question is whether geotargeting can help reduce this delay in taking protective action.

Role of Geotargeted Information in Effectively Communicating Risk

Geotargeting allows an alert or warning message to be tailored according to the nature of the hazard faced at a particular location and the protective action appropriate for someone at that location. Receipt of a geotargeted message also strengthens the recipient's perception that

BOX 1.2
What Is Known About Effective Alerts and Warnings

Several decades of research on public response to alerts and warnings have yielded an understanding of how to best formulate an alert or warning message. Key insights from this work are given below.

Message Content
- *What (guidance).* Exactly what protective action can be taken to best protect health and safety, and how will it be done.
- *When (time).* When protective action should be performed, when it should be completed, and when the threat has ended.
- *Where (location).* What locations are at risk and who should take protective action.
- *Why (hazard and consequences).* What the impending hazard is, what the potential consequences of the hazard are, and how the protective action can reduce harm.
- *Who (source).* Who is providing the message—which should be selected based on who would be viewed as the most credible source for the affected population, often several sources.

Message Style
- *Clarity.* Message is simply worded, free of jargon, and uses words that people can understand.
- *Specificity.* Message provides precise and unambiguous information.
- *Accuracy.* Message contains correct and complete information that is up-to-date, as free from errors as possible, and corrected when better information is known.
- *Certainty.* Message is stated authoritatively and confidently, even when there may be uncertainty about elements of the message content, especially about the protective action to be taken.
- *Consistency.* Message explains any changes from past messages and does not provide conflicting information that may create uncertainty.

Message Delivery
- *Repetition.* Messages may need to be sent multiple times during an emergency.
- *Multiple channels.* Messages should be sent using all available alert and warning systems.

Contextual Factors
- *Status.* Demographic factors, including socioeconomic status, age, ethnicity, and gender.
- *Family and community role.* The roles of individuals within and outside of the family, including family size, child ages, pets, and greater community involvement.
- *Experience.* Personal experiences of prior disaster.
- *Knowledge.* Pre-event knowledge related to the hazard, the protective actions, and the warning system.
- *Environmental cues.* Indicators in one's environments such as rain, wind, or smoke.
- *Social cues.* Seeing or hearing about others taking the recommended protective action.

SOURCE: M. Wood, CSU, presentation at the Workshop on Geotargeted Alerts and Warnings, Washington, D.C., February 2013, drawing on work by Dennis Mileti and John Sorensen.

> **BOX 1.3**
> **Message Variance in Food Recall Experiment**
>
> Message 1: Risk certainty (certain consequences) and expert opinions (cognitive salience).
>
> Message 2: Risk certainty (certain consequences) and concrete experience/personal stories about sickness and death (emotional salience).
>
> Message 3: Expert opinions (cognitive salience), personal stories (emotional salience), ambiguous risk cause (uncertain consequences), multiple behavioral options (multiple efficacious actions).
>
> Message 4: Risk certainty with specific and geotargeted behavioral directives.

he/she is indeed at risk and should take protective action. Finally, because receipt of a geotargeted message can substitute for lengthy descriptions of what areas are at risk, the message content can focus on the hazard and recommended public actions. (The latter is of particular value in systems that strictly limit message length.)

Timothy Sellnow reported a recent experiment testing the different wording for a message on a food safety recall—for example, Do message recipients understand how to protect themselves? and Do they intend to follow the instructions? Box 1.3 lists the various message variables tested. The fourth test message was geotargeted by stating explicitly that the affected food product was distributed in the recipients' area.

Sellnow and his colleagues had two hypotheses: (1) the efficacy scores for behavioral intention, which indicate that a recipient intended to take protective action, would rise after receiving *any* of the four messages, and (2) the geotargeted message (message 4) would result in the highest level of efficacy and behavioral intentions. Hypothesis 1 proved true; recipients' had a greater sense of behavioral intention. This confirms that those who receive a specific message about appropriate response believe they have the skills and intend to take suggested action. Perhaps more related to the workshop topic, hypothesis 2 also proved true; when a message is geotargeted, messages attract more attention, and recipients better understand potential risk. Sellnow stated that they have found this consistently across several experiments that examined geotargeted alerts and warnings. Additional key observations and open questions offered by Sellnow include the following:

- Past work has shown how an optimal message can be communicated in a limited *time*, but further research is needed on whether and how an optimal message can be conveyed in limited *length* (e.g., in WEA).[7]
- Some elements of geotargeting have been shown to improve self-efficacy—that individuals know what the appropriate action is and believe they can complete it—but further work would improve understanding of how other message variables, such as message source or alert type, structure, and delivery systems, affect response.
- The previous experience of message recipients with disasters remains a potentially confounding factor that this work did not control for. What impact might such experience have, and how might it be controlled for in future experiments needs further research?
- Recent research in health communications has shown that differences in recipients' learning styles, or the way in which an individual best acquires or processes new information, may significantly affect their efficacy. How do learning styles affect the efficacy of geotargeted alerts?
- What are the optimal strategies with respect to the number and timing of messages at various points in an emergency? For example, when is it helpful to issue alerts as soon as a potential risk emerges? and Are there circumstances under which alerts are best deferred until the actual event onset?
- What types of hazards need what types of alerts or warnings? How does the prevalence of a particular hazard affect public response and alert needs?

Communicating Geotargeted Information

Michele Wood discussed the three general ways to communicate what areas are at risk: text that names locations or describes the boundaries of the area, pictures that show a map of the area, or geotargeted delivery of alerts or warning messages so that only affected populations will receive the messages.

Text-Defined Geolocation Information

Many alerting systems can transmit only text messages and thus must rely on text to convey geographical information. Possible approaches include the following:

---

[7] B. Reynold and M. Seeger, Crisis and emergency risk communication as an integrative model, *Journal of Health Communications* 10:43-55, 2005.

- *General text:* "Tornado warning in this area until 12:00 PM Mountain Standard Time, take shelter now—NWS." Although the message is short, recipients may not be able to tell whether they have been geotargeted (unless the message explicitly states they were) or whether they are, in fact, located in the affected area.
- *Named locations:* "Radiological hazard warning in Los Angeles, Orange, and Riverside counties until 12:00 AM PST. Take shelter now.—DHS." This message names all the locations affected. However, this type of message is longer than necessary (and possibly longer than can be accommodated in systems that only transmit short messages). Additionally, the boundaries of the area affected by a hazards may not correspond well with the boundaries of cities, counties, or other named places.
- *Descriptive boundaries:* "Fire warning in area bounded by North West Parkway/E-470 on the north, Highway C-470 on the south, Highway 285 on the west, and the eastern boundaries of Adams and Arapahoe counties until 12:00 AM PST. Take shelter now.—NWS." Although this approach makes it possible to define more precisely the area at risk, it does so with greater message length. Furthermore, it can be difficult to identify suitable boundaries that precisely describe the area at risk and that can be easily understood by recipients. Messages that depend on detailed local knowledge are especially difficult for visitors or new residents to interpret.

Using Maps and Images to Geotarget Messages

Maps are, of course, the canonical way to represent geographical regions. Map data are generally readily available, and it can be easy to clearly mark areas at risk through shading or coloring. Adding a marker for the recipient's location further personalizes the information. If the system does not allow the recipient's location to be shown (which relies on capabilities in the device receiving and displaying the message), a recipient may not readily see whether he/she falls within the affected area. Maps can also be difficult to display on devices with small screens (such as cell phones) or with low resolution (such as non-high-definition televisions). Another challenge is that not all message recipients will be adept at interpreting the map.

Geodefined Message Delivery

The third method for geotargeting messages is to target delivery of the messages to affected areas and provide the recipient with an indication that the message was targeted. For example, the message might include the text "If you receive this message, you are at risk," so the recipients understand that they are within the targeted area and are at risk. This strategy will work best if the alerting and warning system can target a sufficiently fine-grained location. (See Chapter 2 for a discussion of these capabilities.) Messages that purport to be geotargeted but are, in fact, delivered well outside the region actually at risk may be ignored because the recipient is not sure whether the message applies to him/her specifically.

Current Research on Message Length, Geographical Information, and Geotargeted Alerts

Brooke Liu, Michele Wood, and their research collaborators, are currently studying the relationships that text alerts, length, and geotargeting have on public response. The research considers how to optimize messages within the length constraints of major text alerting systems—WEA (90 characters), SMS and Twitter (140 to 160 characters), and the instructional field of an EAS message (1,380 characters). The research team is attempting to answer the following questions:

- How can 90-character messages be optimized? How is public response affected by location specificity, the sequence of the location information, and the source of the information?
- What is the relative importance of the key message elements in 1,380-character messages?
- Can these findings be generalized across different types of hazards?
- What is the relative efficacy of different (message length and location specificity) message types?

Because WEA messages are limited to 90 characters, additional research and thought are needed to understand what should be communicated to people first, second, third, and so on. In thinking about this question, Liu developed a matrix of imminent threat hazards and a sequence of protective actions that may need to be taken by an affected population.[8] For example, a primary action in response to a chemical spill might be to shelter in place, and a secondary action might be to get decontaminated. The open question is when geotargeting makes the most sense for which alerts. Liu and her team believe that this is affected by several factors: alert type and format, hazard type and available content,

---

[8] The complete matrix can be found in National Consortium for the Study of Terrorism and Responses to Terrorism, *Hazards and Protective Actions Sequence Matrix: Comprehensive Testing of Imminent Threat Public Messages for Mobile Devices,* 2013, available at http://www.start.umd.edu/start/publications/HazardsAndProtectiveActionsSequenceMatrix.pdf.

and lead time. Additional open research questions identified by Wood and Liu include the following:

- How does geotargeting affect people who are visually impaired?
- How does map literacy affect public response to geotargeted alerts?
- Does precise location language reduce time spent milling? Is this true across hazards?
- What limits does technology place on the ability to geotarget alerts?
- What is the most efficient combination of text, image, and geo-delivery?
- What public education is needed to help the public understand geotargeted alerts and new alerting systems?

Use of Geotargeted Information by Emergency Managers

Ken Rudnicki discussed the capabilities emergency managers have to issue geotargeted alerts and warnings. Each tool has different geotargeting capabilities, as follows:

- EAS provides information to television viewers and radio listeners. Although the messages generally contain descriptions of what areas are at risk, because they reach everyone served by a broadcaster or cable system, they are often disseminated to a large number of people who are not at risk.
- NOAA weather radios can be kept in constant standby mode, have alarms that can wake people who are sleeping, and support those with impaired vision. However, radios must be purchased and kept powered on by the end user.
- Subscription-based SMS text systems are used by many jurisdictions and institutions to send text messages that provide alerts about hazards and other items of interest, such as severe traffic alerts. Although these systems typically allow users to sign up for particular geographical areas of interest, the technology does not support delivery based on the recipient's current location. As opt-in systems, they reach only people who sign up for the service.
- Reverse-dialing systems, currently one of most precise geotargeting systems, allow emergency managers to telephone individual homes within a given geographical region quickly. The population coverage of this technology is diminishing now that an increasing number of households no longer have a landline telephone.
- Third-party smartphone applications have also become more popular. These systems can use the phone location to provide alerts. Similar to subscription SMS text systems, users must opt in by downloading

the application on their phone. Additionally, since a third party manages applications, emergency managers have no understanding of who is receiving the alerts.
- WEA has numerous advantages over prior alerting tools; however, the very short message limits what can be communicated to recipients. While WEA provides county-level geotargeting by cell tower, there is still the potential for bleed over into unaffected areas. Furthermore, as it is a newer system, some practitioners and recipients are not fully aware of the capability or how best to use WEA.

HAZARD TYPE AND GEOTARGETING

The second session focused on the geotargeting needs and challenges for particular hazards. Thomas Cova, University of Utah, discussed wildfire events; Steven M. Becker, Old Dominion University College of Health Sciences, discussed radiological and nuclear incidents; and Peter LaPorte, Washington Metropolitan Area Transit Authority (WMATA), discussed transportation system emergencies.

Wildfire Events

Wildfire events provide significant challenges and opportunities for geotargeting alerts. While some wildfire movement is slow, allowing for proactive evacuation, shifting winds can quickly alter the course of the fire, which results in a need for fast evacuations of very specific areas. Thomas Cova presented the following case studies to illustrate how geotargeted messages might be used in wildfire events:

- The evacuation area for the 2012 High Park Fire in Colorado was described as "CR44H three miles south to just north of Stringtown Gulch Road, the entire Rist Canyon area, CR27E to Stove Prairie and south to Davis Ranch Road and Whale Rock Road." Such a description poses two challenges: first, it cannot be understood by someone not very familiar with the area, and second, it turns out that this description does not, in fact, define a closed polygon.
- The evacuation zone established for the 2007 Angora Fire provides an example of how a very small and precise evacuation zone can be established using polygons. The evacuation order had 100 percent compliance for two reasons: (1) the warned area was small and highly targeted and (2) the flames from the rapidly moving fire provided environmental cues that action needed to be taken. The Angora fire also illustrates some of the complexities posed by geography. Because many of the roads in the neighborhood led into the fire's path, many evacuees had to flee on foot.

- In the case of the 2000 Cerro Grande Fire in Los Alamos, New Mexico, emergency responders had as much as a week to carefully plan and carry out evacuations. They were also aided by very detailed emergency plans already in place, given the proximity of the U.S. Department of Energy laboratory. As the time to evacuate approached, houses in each designated zone were notified using a reverse 911 telephone dialing system of the need to evacuate and given instructions on when and by which roadways they should evacuate. The evacuation was completed quite swiftly, in 2.5 hours, and with almost 100 percent compliance.

Geotargeting could also provide an opportunity for time-sequenced changes in the recommended protective action for those who have not followed previous instructions. For example, the first alert within an area would instruct recipients to evacuate, a second would recommend that remaining residents shelter in place, a third would recommend sheltering in a refuge, and a fourth would recommend finding a safe area (a protective structure or body of water).[9]

Radiological and Nuclear Incidents

There are different types of radiological and nuclear incident types, including the unintentional release of radioactive material from a container break, nuclear accident at a nuclear power plant or nuclear fuel processing site, and an intentional radiological or nuclear incident, either a radiological dispersal device (i.e., dirty bomb) or a nuclear explosion of an improvised device. For each type of event, the right protective action information needs to be disseminated to the right people as quickly as possible. Steven Becker discussed the substantial challenges of messaging during radiological and nuclear incidents, including the following:

- Uncertainties in determining the composition and direction of radioactive plumes, changing conditions that require frequent changes in information, and, when terrorism is suspected, concern about additional attacks.
- The potential for widespread fear, profound sense of vulnerability, and a continuing sense of alarm and dread, which research has shown to be associated with emergencies involving radiation. Fatalism is particularly high with radiological events, and reportedly especially high within

---

[9] T.J. Cova, F. Drews, L. Siebeneck, and A. Musters, Protective actions in wildfires: Evacuate or shelter-in-place? *Natural Hazards Review* 10(4):154-162, 2009.

minority populations.[10] There is a high propensity to flee in these situations, and compliance with shelter-in-place instructions may be poor.
- The complexity of and unfamiliarity with radiation-related concepts and terms.

Although there are messages, templates, and questions and answers that have been developed and vetted for technical accuracy and efficacy, less is known about how these messages can be communicated in a shorter format. There is also a large demand by affected populations for graphics of the affected area, especially in regard to plume maps. Messages also need to be tested and developed that can address the problem of fatalism.

Fukushima Dai-ichi Nuclear Accident

During the March 2011 Japan Earthquake, the Fukushima Dai-ichi Nuclear Generating Station, one of the largest such installations in the world, was severely damaged. It would become clear that large amounts of radioactive material had been released, that there were threats of additional releases, and that a large area needed to be evacuated. Although plans had been created for areas immediately surrounding the plant, the evacuation zone became much larger, complicating evacuation planning and execution. The event provided valuable lessons in communicating during nuclear and radiological incidents:

- Evacuation orders for radiological events need to include information about direction of the plume so that people can take appropriate action. This information was not provided because scientists felt that the plume model was not fully developed and because emergency managers feared that people would panic if they were provided incomplete information. Unfortunately, the decision to withhold information resulted in people in some communities actually evacuating into the path of the plume.
- Emergency plans need to handle cases where the infrastructure is degraded. At Fukushima, individual emergency plans had been developed for earthquakes, tsunamis, and nuclear accidents. But when all three occurred nearly simultaneously, many of the shelters or evacuation routes were no longer usable. Similar concerns apply to the communication infrastructure needed to alert the population and coordinate emergency response.

---

[10] S.M. Becker, Emergency communication and information issues in terrorist events involving radioactive materials, *Biosecurity and Bioterrorism* 2(3):195-207, 2004.

- Messaging strategies and preparedness education need to include methods for the public to recognize genuine communications and identify misinformation. Various credible-looking but false messages were sent to people in the affected area. This resulted in people taking potentially unsafe or useless actions, such as purchasing large amounts of iodized salt in the false belief that this would provide useful protection against radioactive iodine isotopes.[11]

Public Mass Transportation Systems

Peter LaPorte, director of emergency management for WMATA, discussed the challenges of communicating with riders of public transportation systems. Most messages to the public relate to system delays or temporary closures, while a much smaller number are issued when passengers are directly affected by breakdowns or other problems. Many alerts are to notify riders about delays owing to train, track, or signal problems that require trains moving in opposite directions to share a single track temporarily. Alerts may also cover scenarios that have a larger and longer impact on riders, such as a broken rail or a station fire. In these incidents, it is important to convey what the public should expect in terms of delays or closures. Some scenarios, such as an individual on the tracks, pose a different challenge because authorities are trying to understand a complicated and potentially serious situation at the same time that the public and the media also want to understand what happened. Did someone jump? Was someone pushed? Did someone fall?

Messages are delivered through multiple channels. Within the system, WMATA can use electronic message boards and public address systems in the stations and public address systems on board the trains. WMATA also provides text message and email alerting and allows customers to subscribe to alerts by line and time of day. WMATA alerts regarding major delays are also picked up by alerting systems operated by local jurisdictions and the media. In some scenarios, such as when a train is disabled between stations, responders may be needed on the scene to directly communicate appropriate action and to avoid unsafe actions by passengers.

The various capabilities for alerting the public all have certain limitations. For example, cellular coverage in the underground portion of the system is not complete, although a congressional mandate to provide full

---

[11] Discussion of ways to identify and correct misinformation during disasters is discussed in depth in the previous report, National Research Council, *Public Response to Alerts and Warnings Using Social Media: Report of a Workshop on Current Knowledge and Research Gaps*, The National Academies Press, Washington, D.C., 2013.

coverage is expected by the end of 2014. In addition, public address systems in the trains and stations may not be audible to all riders.

WMATA's alerting process also includes alerting and contacting local-jurisdiction emergency or transportation staff. WMATA works closely with the federal Office of Personnel Management (OPM) during severe weather events to consider the impact of service disruptions on the federal workforce; many private organizations also follow the OPM's lead. WMATA also works with local trade organizations to communicate with local workforces. Responding to emergencies within the system involves close cooperation and careful coordination with county and city authorities and first responders in Maryland, Virginia, and the District of Columbia.

DATA SECURITY AND PRIVACY CHALLENGES

Panelists Patrick McDaniel, Pennsylvania State University, Marc Armstrong, University of Iowa, Darrell Ernst, private consultant, and Kevin Pomfret, Centre for Spatial Law and Policy, examined data security and privacy concerns associated with geotargeted alerts and warnings.

Defining Security and Privacy

Security was defined by Patrick McDaniel as when a device, system, or service acts as expected. When a system is secure, information is legitimate and unmodified, only authorized parties can participate, access is available, and, where needed, secrecy (not to be confused with privacy) is available. In the context of alert and warning systems, security concerns include ensuring that only those who are authorized can send alerts, that messages cannot be modified or forged to misinform the public, and that messages cannot be suppressed through damage to or overload of the delivery system.

Privacy, as defined by McDaniel, means that an individual controls what information about that individual is exposed and to whom, when, and for what purpose. Privacy is often a negotiated and constantly evolving term. The issue is mostly of control, not exposure, and privacy starts with informed consent. Users often trade some information for a benefit; they need to understand this implicit transaction. If a company collects information for a benefit then uses that information for an undisclosed purpose, that is a violation of privacy. While the majority of privacy conversations involve knowing a user's physical location, collection of location over time can provide a very detailed outlook of an individual. Users often have different reactions to a service provider having location information versus the government having the same information.[12] The design of an alert and warning system is influenced by both security and privacy, and retrofitting a system to consider these concerns is difficult.

A geotargeting service might adhere to some of the following basic, demonstrable, and testable principles and standards to provide security and privacy:

- All alerts must be delivered without modification.
- Only authorized parties are able to transmit alerts.
- All alerts must be received within a specified period. An example of this is E911, which requires that a location of a user within 300 meters can be found within 6 minutes.
- Alerts must be delivered only to the designated area.
- Alerts must not substantially affect the infrastructure over which they are transmitted.
- Users must be able to control or know what personal information is exposed to the service provider and how the service provider will use the information.

Defining Geoprivacy

Marc Armstrong defined geoprivacy as an individual being secure from unwanted observations and tracking. Currently there are requirements involving location data and mobile devices. This includes E-911, which requires that mobile devices periodically report locations based on Global Positioning System (GPS), cellular towers, or Wi-Fi locations. Similar location data are often shared with third-party application developers, who may also then share the data with others. Risk associated with this behavior includes further disclosure of information, consumer tracking, identity theft, threats to physical safety, and surveillance.

There are some techniques to preserve geoprivacy while still providing location-based services. One is areal masking, in which location information is sent only when someone enters a particular region. Travel within the region would not need to be known. However, this would delay the immediate arrival of geotargeted notifications to the end uses. Current work is being done that highlights these masking techniques and a recently founded company, iGeoLoqi, has developed tools that incorporate some of these techniques, which provide location information and broadcast push notifications, and individuals trigger the messages by entering, staying, or leaving a location.

---

[12] U.S. Government Accountability Officer, Mobile Device Location Data, 2012, available at http://www.gao.gov/assets/650/648044.pdf.

Legal Issues Around Geotargeted Alerts

Kevin Pomfret defined spatial law as the set of legal issues associated with geospatial technology and the collection, use, and transfer of location information and other types of spatial data. Spatial law involves a range of issues, including privacy, intellectual property and licensing, data quality and liability, national security, open records, the Freedom of Information Act, and other government regulations. Location information creates challenges distinct from those created by traditionally protected personal information (such as social security numbers and financial information). For example, someone who does not have other information about a person's identity can, nonetheless, readily infer where the person works and where the person lives. In addition, location data are generated differently from other sorts of personal information, making the definition of "public"—long an analytical tool used in assessing privacy and other impacts—a challenge. For example, individuals move (change location) through public spaces, and most government privacy policies have exceptions for "publicly available information."

Even as compelling new applications for location information are being developed and deployed, privacy advocates urge careful consideration of how this information might be used and how to protect the privacy of users. Media scrutiny is having an impact on this discussion. For example, due to media coverage, organizations are pulling back the use of location tracking tools. Short Pump Mall in Richmond, Virginia, decided against using a mobile application after media scrutiny criticizing its privacy implications,[13] and a New York community stopped using publicly available aerial imagery to identify unregistered and untaxed pools. Additionally, Congress has considered several pieces of legislation addressing location privacy, and regulatory authorities are examining ways to address this issue. However, geolocation information is often just part of a longer list of issues surrounding privacy.

Previously enacted privacy legislation—such as the Commercial Privacy Bill of Rights Act of 2011, BEST Practices ACT, Do Not Track Me On-line Act, Consumer Privacy Protection Act of 2011, and an update to Children's Online Privacy Protection Act—may provide a foundation for future legislation protecting location information. Many of these laws are

---

[13] *Richmond Times-Dispatch*, Town Center's Monitoring Policy Creates Backlash, posted November 26, 2011, available at http://www.timesdispatch.com/business/town-center-s-monitoring-policy-creates-backlash/article_7b6bd70f-7bc8-5564-a417-7856266079f5.html.

based on the principles of fair information practices (FIP).[14] Elements of FIP include notice and transparency, consent and use limitations, access and participation, integrity and security, and enforcement and accountability. However, there are challenges in applying FIP to geolocation information: How will "precise geolocation information" be defined? How do you provide adequate notice on a mobile device? In which contexts are opt in or opt out the appropriate forms of consent? What is a consumer's right to have geolocation information corrected or deleted? and How long should geolocation information be permitted to be stored?

The Federal Trade Commission (FTC) is examining privacy-protecting practices and has broad authority under Section 5 of the Federal Trade Commission Act. These actions include bringing action against companies that do not comply with their own privacy policies and against companies that do not adequately protect personally identifiable or other sensitive personal information. A December 2012 FTC report, "Protecting Consumer Privacy in an Era of Rapid Change—A Proposed Framework for Business and Policymakers" provides a list of the FTC's best practices for mobile privacy (summarized in Box 1.4).[15]

Technologies That Support Privacy

In 1992, Darrell Ernst began developing a technology to geotarget alerts while preserving privacy. This technology would send an alert to a wide area, and a location-aware device would determine if the alert was relevant to individual users. The technology was originally developed to aid in risk communication to military personal (information regarding detected missile launches is broadcast to specifically programmed handheld devices carried by personnel in the field). If a device received such a message and determined it was in the warning area, it would notify the user. A general application of the technology could allow for more privacy-sensitive alerting. Since the device itself assesses and filters messages, the authorities sending the messages do not need the ability to track where end users are. Receivers for this type of alert broadcast could be placed in a wide variety of devices, including in-home appliances that are programmed with their static location. Called GEOcast, SquareLoop

---

[14] Fair Information Practicies (FIP) were developed by the Federal Trade Commission in the mid-1990s in response to concerns of how online entities collect and use personal information. Current FIP principles can be found at http://www.ftc.gov/reports/privacy3/fairinfo.shtm.

[15] Federal Trade Commission, *Mobile Privacy Disclosures: Building Trust Through Transparency*, Staff Report, February 2013, available at www.ftc.gov/os/2013/02/130201mobile privacyreport.pdf.

> **BOX 1.4**
> **The Federal Trade Commission's Best Practices for Mobile Privacy**
>
> Provide timely privacy disclosures to consumers and obtain their explicit consent before allowing apps to access and collect certain sensitive data and content;
> Consider developing and implementing a visual "dashboard" that displays for consumers the types of data accessed and collected by apps;
> Consider designing, testing, and implementing intuitive and simple icons to depict certain app privacy practices;
> Implement and enforce contractual obligations for, and promote best practices and educational information to, app developers that address mobile privacy;
> Consider providing consumers with clear disclosures about the extent of prerelease review and postrelease compliance checks that platforms undertake for apps that can be downloaded from the platform; and
> Consider offering a Do Not Track (DNT) option.
>
> SOURCE: Federal Trade Commission, *Mobile Privacy Disclosures: Building Trust Through Transparency,* Staff Report, February 2013, available at www.ftc.gov/os/2013/02/130201mobileprivacyreport.pdf.

licensed this technology and continues to further its development for alerts and warnings.[16]

Workshop attendees noted that some current alerting systems use similar privacy-protecting methodology. For example, with WEA, cellular towers determine if they are within a given alerting area and send alerts to all phones within the designated area. Much of the second-day discussion at the workshop focused on technologies, such as the one described above, that would support better localization by the device itself. If a device had more precise knowledge of its location, this information would not necessarily need to be shared with alerting authorities.

---

[16] See http://www.squareloop.com/.

2

Technologies and Tools for Geotargeted Alerts and Warnings

Several recent innovations or capabilities under development provide some of the necessary ingredients for an end-to-end, all-hazards warning system that fully exploits geographical information:

- The Common Alerting Protocol (CAP) standard for formatting alerts includes geographical locations by Federal Information Processing System (FIPS) code or vertices of polygons to define affected regions along with information about the source and nature of the alert and the action to be taken.
- Cellular phones and other mobile devices "know" where they are located (at a minimum using mandated E911 location capabilities and, increasingly, using embedded Global Positioning System (GPS) receivers and other location information such as nearby wireless access [Wi-Fi] sites) and increasingly possess considerable processing power, high-resolution displays, and the like. More generally, computing devices, such as laptops, desktops, and cable set top boxes, either can establish their location or can easily be outfitted to determine such information using one or more of the approaches listed above. Wired devices can also use knowledge about the physical location of the networks to which they are attached to establish their location. Applications either built in or installed on these devices can be used to receive and present targeted alerts and warnings. Importantly, even if the systems designed to transmit alerts/warnings cannot precisely send messages only to the desired set of recipients, the receiving device can use knowledge of its location, together with

geographical information coded in the message, to deliver messages only to someone at the specified location.
- Tools for geotargeting at various resolutions are becoming increasingly available, and these have been adopted by the advertising industry. The alerting community may be able to adopt or adapt capabilities being developed and used for advertising.

Much of the second day of the workshop focused on new and emerging technologies and tools for determining location and disseminating geotargeted alerts and warnings.

CONTINUING OPPORTUNITIES FOR USING TRADITIONAL TECHNOLOGIES FOR GEOTARGETED ALERTS AND LESSONS FOR THE USE OF NEW TECHNOLOGIES

Although discussion at the workshop tended to focus on new technologies, particularly mobile devices, several presentations examined how older technologies can be used in ways that provide enhanced geotargeting capabilities. Rick Wimberly, Galain Solutions, examined reverse-dialing alerts; John Kean, NPR Labs, discussed innovations in radio broadcast; Bruce Thomas, Midland Radio Corporation, examined weather radios; and Ron Boyer, Boyer Broadband, discussed alerting over cable television systems.

Telephone Alerting

Reverse-dialing alerts allow for officials to auto-dial landlines, or mobile numbers of registered users, within a certain area and play a prerecorded alert. As noted by workshop participant Ken Rudnicki, City of Fairfax, Virginia, reverse-dialing systems are currently one of the better tools emergency managers have to provide geotargeted alerting. However, the system still has several challenges.[1] The challenges, as discussed by Rick Wimberly, include the following:

- When large sets of numbers are dialed, this can overwhelm local phone switches and cause calls to be dropped.
- The significant decrease in the number of households with landlines reduces the reach of these systems. Reverse-dialing systems can

---

[1] In a 2012 article, Rick Wimberly examined the shortcoming of these systems during the 2012 Colorado wildfire where approximately 25,000 of the 118,000 reverse-dialing alerts were not delivered. See R. Wimberly, Flawed delivery: Do alert notifications fail to live up to expectations, *Emergency Management Magazine* 5(7):22-35, 2012.

reach mobile devices, but this requires subscribers to register their phone numbers. Despite communities aggressively encouraging people to register, registration rates for mobile subscribers across the country are still well below 10 percent.

- Reverse-dialing systems are not particularly effective at delivering messages to those with disabilities.
- Reverse-dialing systems are expensive, and local jurisdictions may not be in a position to purchase or modernize a system.

Radio Broadcast Technologies

NPR Labs, a small, self-supported broadcast technology research and development outfit operated by National Public Radio, is currently examining the use of two new technologies that may benefit alerting: broadcast repeaters and the use of the radio broadcast system (RBDS).

NPR Labs partnered with Geo-Broadcast Solutions (GBS) to examine the performance and use of GBS technologies known as ZoneCasting and MaxCasting. In both technologies, a group of synchronous repeaters repeats the signal of the primary station using lower power and transmitter heights. In MaxCasting, the nodes are time-aligned to the primary transmitter to reinforce or extend coverage. In ZoneCasting, the individual nodes can be used to send distinct programming to different locations. Figure 2.1 demonstrates how these tools can expand coverage of a radio station and also provide separate coverage by zone.

John Kean discussed how both tools support alerting: first, they extend the reach of radio alerts to communities currently poorly served by single radio transmitters; and, second, by supporting distinct programming by different nodes, they enable geotargeting of alert and warning messages. Although they require new equipment on the part of the broadcaster, they have the advantage of requiring no new equipment for the public.

NPR Labs is also working to demonstrate the use of RBDS to reach at-risk populations, including those with hearing impairments. RBDS is a standard to embed small amounts of digital information in conventional radio broadcasts that almost all FM stations are capable of supporting. It is currently used most often to transmit and display song or other program information and is commonly found in automobile radios. One of the objectives of the NPR Lab project is to experiment with using RBDS to send text information using household receivers to people with hearing impairments to explore how effectively this technology would reach this large segment of the public.

TECHNOLOGIES AND TOOLS FOR GEOTARGETED ALERTS AND WARNINGS   27

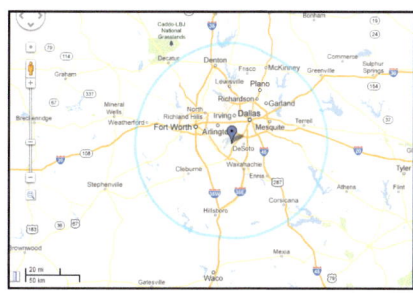

(a) KVIL-FM is a class C FM station, the largest classification for stations. The station has coverage of approximately 24,000 square miles (with a coverage radius of 88 kilometers). A smooth circle usually represents this coverage; however, coverage is not that consistent.
Circle added by NPR Labs, map copyright 2013 Google.

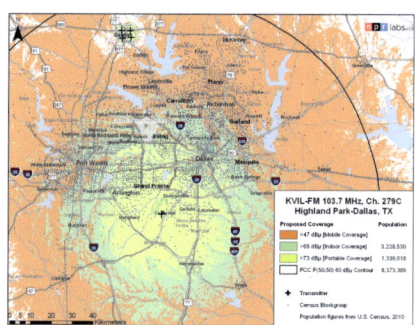

(b) Broadcast is a terrestrial signal, so the signal strength is impacted by terrain. Coverage is also lost due to less efficient antennas and building penetration. For this station, coverage should be approximately 6,373,000 people. When terrain sensitivity and indoor penetration are factored in, coverage shrinks to about 3,173,000 people.

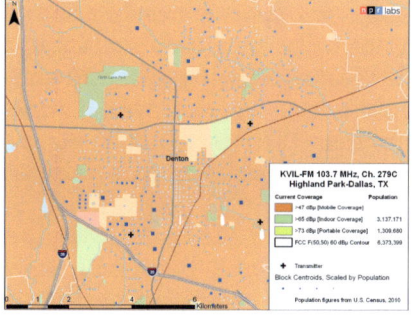

(c) Denton, Texas, falls within the station's coverage area; however, almost the entire town sits within an area where there is little to no coverage.

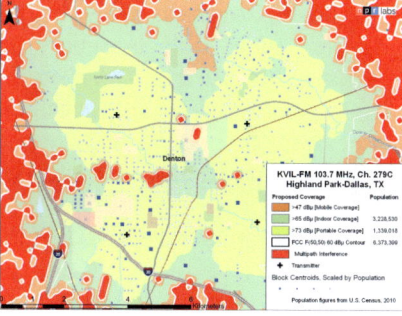

(d) When ZoneCast nodes are added to Denton, the signal can reach indoors and also allows for special announcements to separate areas.

FIGURE 2.1 Radio broadcast coverage of KVIL-FM, Dallas, Texas.
NOTE: Radio coverage in images b-d is indicated by shading; darker areas have basic coverage, and lighter areas have increased coverage, including indoors and in previously terrain-blocked areas. SOURCE: John Kean, NPR Labs, presentation at the Workshop on Geotargeted Alerts and Warnings, Washington, D.C., February 2013.

NOAA Weather Radio

The National Oceanic and Atmospheric Administration (NOAA) Weather Radio (NWR) was originally developed in the 1950s and 1960s to provide weather observations and forecasts to those in flight or at sea. In 1974, an outbreak of 146 tornadoes within a single day spurred the expansion of the service. Bruce Thomas noted that today, more than 1,000 broadcast transmitters provide coverage across the United States and its territories.

NWR uses the Specific Area Message Encoding (SAME) standard to geotarget its alerts. Adopted in 1988, SAME was the first geotargeted alerting standard and remains an important foundation for geotargeted alerts to this day. SAME allows NWR to target at the FIPS code level. This generally means at the county level (or equivalent geographical area); however, large cities located within counties may have their own unique FIPS and SAME codes. Additionally, high-risk areas may have a unique SAME, such as areas around a nuclear power plant. More precise geotargeting may be possible by adjusting the first digit of the SAME code.

Cable Television

Boyer discussed current capabilities and opportunities for enhanced alerting over cable networks (more formally known as multichannel video programming distributors or MVPDs). Currently, cable providers are required to distribute presidential alerts, test their alert system weekly, and monitor two Emergency Alert System (EAS) origination sources. Many operators support additional alerting capabilities on a voluntary basis.

Today, an alert received from EAS is distributed to all the subscribers to a cable system, even though they may live outside the specific region that is the subject of the alert. In principle, modifying cable boxes to know their location and filter messages accordingly could provide better geotargeting. Boyer explained that this is not entirely straightforward since cable boxes have been designed chiefly to decode video content, were not designed to be location-aware, and the systems that would be needed to link a subscriber's cable box to the subscriber's address in the MVPD's billing or operational systems do not exist today.

A second option would be to determine location via service nodes within the cable system. However, these do not necessary follow the geopolitical boundaries commonly used to geotarget alerts. Boyer noted that, as a result, adding enhanced geotargeting to MVPDs will most likely require enhancements to the entire networks, not just a single network element.

TECHNOLOGIES FOR GEOTARGETING ALERTS OVER THE INTERNET

Although television is still the primary source by which the public receives information about disasters, this is rapidly changing as individuals spend more time using the Internet for infotainment. Richard Barnes, BBN Technologies, discussed the challenges of alerting over the Internet, and Hisham Kassab, MobiLaps, discussed how alerts and warnings can be introduced into streaming video content.

In principle, alerting over the Internet appears to be a straightforward task requiring, essentially, the delivery of a suitably formatted document that the Internet-connected device can render, something that is done trillions of times a day. However, this would require that the location of the Internet-connected device can be established by some combination of the device itself and the network and that a device either monitors for alert information or the alert can be placed in an information source that is already being monitored by the device.

Geotargeting Using Internet Protocol

Richard Barnes explained that geolocating based on Internet protocol (IP) address is very limited and generally relies on privately managed databases that match IP addresses to physical addresses. IPV6 has extra space, a 128-bit address, and many hoped that this extra space could be used to insert geolocation information and allow for better IP topology-based geolocation. However, IP addresses are assigned based on the network topology, not physical geography. In addition, IPV6 might make IP-based geolocation more difficult because the larger address space may make it more difficult to complete network traces used to determine location. Essentially, IPV6 has the same geotargeting capabilities as IPV4, where geotargeting using network tracing can be done in a metro area. (Law enforcement can obtain additional information on a geolocation of an IP address from a service provider with a subpoena.)

Another method for geotargeting would be to incorporate alert and geographical information in an information source that a user already frequently monitors. For example, an alert could be sent to Facebook, which would then send the alert to all of its subscribers in Nebraska. This bounds delivery of alerts to specific channels, and these platforms have their own challenges in accurately geolocating their subscribers and geotargeting messages.[2]

Alerting Over Streaming Video

Increasingly, Hisham Kassab noted, Internet-delivered services, such as Hulu Plus, Netflix, and YouTube, that stream video over the Internet are starting to be used in place of traditional broadcast and cable television for which alerting systems already exist. There are a variety of ways to deliver and display alerts on Internet-connected devices like computers, tablets, or game consoles, including streaming video services and the applications that display them. With the right modifications, these services can be used to receive and display geotargeted alerts delivered or triggered by the video stream and displayed by the application used to view the content.

There are four steps to streaming video, content creation (e.g., Warner Brothers films a television episode), content provision (e.g., Hulu licenses content and makes it available to its subscribers), content transmission (e.g., the viewer streams content over a Comcast broadband connection), and content presentation (an application running on a device displays the content). It obviously is not practical to insert alert information at the content creation step, but alerts could potentially be inserted at any of the other three steps, each with a differing capability to geotarget alerts:

- *Content provider.* Two possible mechanisms could be used to geotarget at the content provider level: the billing address, as a proxy for location, or an IP address. Both of these have limited accuracy because a person may be accessing content away from their home, and IP address databases are not always accurate, as discussed in the previous section.
- *Content transmission.* ISPs may be able to use their knowledge about network topography to determine the physical location of their users.
- *Content presentation.* End-user devices, most notably tablets and smart phones, often have some information on their location. An alert could be inserted into the video stream, and once the alert reaches the device, an application could use its location to determine if the information is relevant.

---

[2] For discussion of the challenges of alerting over Facebook and other social media platforms, see National Research Council, *Public Response to Alerts and Warnings Using Social Media: Report of a Workshop on Current Knowledge and Research Gaps*, The National Academies Press, Washington, D.C., 2013.

MOBILE DEVICE LOCATION DETERMINATION CAPABILITIES

There are several ways in which a mobile user's location might be determined. Farshid Alizadeh, Skyhook Wireless, Larry Dodds, TruePosition, and Ayman Naguib, Qualcomm, presented potential technologies for determining mobile device location.

Mobile Location Determination Using Wi-Fi Access Points

Traditionally, mobile device location has been established using two methods: GPS, which is fairly accurate but only works outside and takes significant time to obtain a location fix and cellular tower triangulation, which has comparatively poor accuracy but provides a faster location fix. Skyhook incorporates a third source, Wi-Fi access point signals. The technology works by matching access point and cell tower signals to a proprietary location database. The technology has a median accuracy of 20 to 40 meters and can return a location fix within a few seconds, and currently, Skyhook has almost 400 million access points in its database.

A unique challenge to using Wi-Fi access points is their dynamism. Wi-Fi access points are under varying people's control and are dynamic—being moved, being removed, and being added. To compensate for this, Skyhook relies on a huge redundancy in these access points. Additionally, the location database is updated frequently. For example, when a user sends a snapshot of surrounding access points to Skyhook servers, not only does the system return a location, but it also incorporates those data into the database to calibrate the location of the access points.

Mobile Location Determination Using Television Broadcast Signals

TrueFix TV Positioning, developed by TruePosition, uses over-the-air (OTA) television broadcast signals to determine mobile device location. Similar to using Wi-Fi signal strength to determine location, these devices search for local broadcast signals and use the signal arrival times, which are directly proportional to the distance to the transmitter, to determine the location. A key benefit of this technology is that it can be used in indoor and urban environments where GPS signals are not able to penetrate building structures. According to Dodds, about 95 percent of the U.S. population lives in areas where the OTA coverage is sufficient to determine position. However, to fully use OTA for mobile device locating, a television-band receiver would need to be added to handsets.

Mobile Location Determination Using Uplink Time Difference

UpLink Time Difference of Arrival (U-TDOA), also developed by TruePosition, uses the time of arrival of signals at multiple cellular towers. Measurements are made using devices that are 1,000 times more sensitive than traditional base stations and are located at or near cellular towers. Several measurements are sent to a central node that calculates the location with fairly high accuracy, typically within a 50-m radius of the correct position. The technology is widely used today by cellular companies to provide E-911 services.

Indoor Geolocation of Mobile Devices

Indoor geolocation is more difficult because GPS and other signals used to determine location do not readily penetrate building structures, and because a significantly higher accuracy is needed indoors. Naguib noted that a 10-meter error while driving is barely noticeable, but it would be problematic for someone navigating through a building.

Qualcomm is developing an indoor positioning technique that uses three additional data sources to determine indoor positioning:

- Wi-Fi measurements;
- Building maps, which provide additional information on what locations are viable (places the receiver could not be located, such as within a wall) and routing information (transitions that are impossible, such as crossing through walls); and
- Sensors on the phone such as accelerometers, gyroscopes, and compasses that can be used as an inertial navigation system by which a prediction is made based on relative motion of the device from its previous position.

CURRENT AND FUTURE TECHNOLOGIES FOR GEOTARGETING ALERTS TO MOBILE DEVICES

As described in Chapter 1, WEA provides limited capabilities for geotargeting alerts to mobile users. New technologies and innovations may provide additional capabilities for alerting and more narrowly defined geotargeting. George Percivall, Open Geospatial Consortium, discussed the use of Short Message Service (SMS) to report geotargeted information; J.T. Johnson, Weather Decision Technologies, described a third-party application for geotargeted weather alerts; and John Davis, Sprint, discussed possible approaches to enhancing the geotargeting capabilities of WEA.

Geotargeting of SMS

SMS is almost universally supported on mobile phones and widely used to send and receive text messages. George Percivall discussed the Open Geospatial Consortium's (OGC's) work to develop a standard, Open GeoSMS, for representing location information in SMS messages. The location can be displayed in a mapping tool or used to retrieve satellite images or other information about the location. Open GeoSMS was used in the mobile phone app Find Me Maybe, which was developed and deployed for limited use during Hurricane Sandy in 2012. The tool also subscribed to the FEMA SMS alert service.

Third-Party Application Capabilities

Using an application on a mobile device and location information from that device is another method for geotargeting alerts. One such application is iMap Weather Radio, which was developed by Weather Decision Technologies (WDT). iMap Weather Radio communicates NWS alerts to the public with the goal of providing some of the key features of NWR (e.g., always on and provides an alert tone that awakens users) on smartphones, which are much more widely deployed. J.T. Johnson noted that iMap Weather Radio offers the following features:

- Phone wakes up automatically with alerts;
- Alerts are for current location and for additional saved locations;
- Interactive maps provide radar view, alert polygons, and phone location; and
- Text and text-to-speech of alerts reaches drivers or those with visual impairments.

To improve accuracy and continuity, WDT uses triple redundant feeds from the NWS and clusters of redundant computers; the data center has an uptime of 99 percent. WDT has also begun working to incorporate CAP EAS into iMap Weather Radio by using the FM channel. WDT also works with local media so that an individual receiving an alert can then watch the local television news directly within the application, either as real-time content or prerecorded material.

Carrier Geotargeting of WEA

As discussed in Chapter 1, targeting methods used by carriers to deliver WEA vary. Geotargeting an alert to a small, defined area is the ultimate goal in the alerting community. John Davis highlighted challenges and opportunities for better geotargeting of mobile devices. These include the following:

- The current deployment of the long-term evolution (LTE)[3] standard may provide a partial solution. With LTE, cellular IDs, which are used to determine if a tower is within the alerted area, are assigned to individual antennas on each tower, rather than the tower as a whole. This may allow for tighter sectioning of geographical regions within the tower's signal.
- The use of GPS is a possibility for determining a mobile device's location, but may pose a challenge if the GPS initiates a request to carrier networks to request location information. Davis noted that this may create congestion in the network and cause its failure. Furthermore, Davis reiterated that GPS inside buildings or on subways is a challenge, and commercial needs will drive the development of tools to allow this.
- Alerting systems may not need to do the geotargeting. A phone's position may be determined by a combination of technologies, as described in previous sections, and then the phone can determine if an alert applies to its location.
- Another challenge is that geotargeting capabilities vary across carriers and devices and are an area of extreme competition between carriers. This creates barriers to discussions across organizations.
- The size of the message is probably the greatest hindrance. Davis explained that the biggest gain in encouraging appropriate public response, with the least impact on networks and devices, is the modest expansion of message information. An option to do this is to allow pagination of a message—that is, a series of messages that together provide the full alert text.

---

[3] Long-term evolution is often marketed as 4G LTE.

3

Research Needs and Implementation Challenges

The following sections outline research opportunities and associated implementation challenges identified by attendees of the workshop. The opportunities and challenges compiled here by the committee from presentations and discussions at the workshop do not reflect a consensus of the committee or the workshop participants, nor are they intended to be a comprehensive list of research questions.

FACILITATING AND IMPROVING PUBLIC RESPONSE

While there are decades of research on public response to traditional alert and warning technologies, less is known about how the public may respond when new technologies are used for alerts and warnings. As noted in Chapter 1, much research is currently being done to answer several of the questions around public response, including the following:

- What are the opportunities for optimizing message content, structure, and alert delivery systems?
- To what extent could more precise geotargeted alerts and warnings reduce mortality, morbidity, human suffering, and other costs associated with disasters?
- How can better communication of locations at risk and appropriate protective actions for each location help reduce delay in recipients' taking protective actions, for example, because less time is spent seeking additional or confirming information or milling?

• What are opt-out rates and causes? Would more tailored opt-out options reduce the opt-out rate? For example, would allowing people to opt in or out of specific categories or sources of warnings help? If so, is there an optimal level of granularity for such selections? How can people be encouraged to opt back in?

• For a given alerting system, what is the most effective repetition frequency for messages and the most effective update interval during an extended emergency? Does this depend on the nature of the hazard or message?

• How can one determine the effectiveness of the alerts or warnings issued during an event? Can the effectiveness of particular messages or systems be measured directly? If not, are there indirect indications that can be used to gauge effectiveness?

VALUE OF GEOTARGETED INFORMATION

Geotargeted information can be presented in several ways: for example, text that includes a place name (e.g., city, county, or zip code), a map that clearly delineates the affected area, a map that shows the affected area and includes the location of the message recipient, or some combination of these. Open research questions regarding the display and presentation of geotargeted information include the following:

• How would using multiple approaches to communicate geotargeted alerts and warnings enhance personalization of risk and subsequent public protective action response? What combinations of text and maps would best motivate recipients to take protective action?

• Does the content of the message—for example, the hazard being warned about or the protective action being urged—play a role in which presentation method is best?

• Given that targeting and sending messages to recipients in unnecessarily large geographic areas can lead to frustration and opting out, what is the most effective size of geotargeting boundaries?

• If maps are incorporated into alerts, what is the most effective method to represent, transmit, and display them? For example, should the maps be represented as raster or vector images? What image size and resolution are best? What level of compression is needed? Should messages contain the maps themselves or just links (e.g., URLs) to the maps?

• How will map literacy in the general populace affect public response? Will map literacy need to be incorporated into disaster education?

• What are the most effective design and visualization principles for geotargeted alerts? Which map symbols, scales, labels, and point of interests would be most effective? Should a standardized set of map symbols for displaying various disaster scenarios be developed?

- What can be learned and applied to alerting tools by examining the use and innovations of mobile and location-based advertising?
- How can geotargeted messages be made more accessible, especially for individuals with disabilities?

DEVELOPING AND DEPLOYING TECHNOLOGY

Communication technologies have greatly evolved in the past decade and will continue to evolve. While much of this innovation is driven by non-disaster response sectors, these technologies will inevitably shift how best to communicate with the public during disasters. Research gaps concerning the use of new technologies include:

- How can new technologies developed by the private sector be adapted quickly and effectively for delivering geotargeted alerts and warnings? What is the role of the third-party developers (e.g., smartphone applications) in delivering geotargeted alerts and warnings?
- What respective roles will special-purpose alert and warning systems (e.g., WEA or NWR) and general-purpose messaging systems (e.g., SMS or social media) play in delivering alerts and warnings to the public? What are the benefits and challenges of each type of alert system? How do they complement each other?
- What legal, regulatory, technology standard, or other barriers stand in the way of rapidly deploying new technology for delivering alerts and warnings?
- How can the gap between what is understood about public response and the technology available for delivering alerts be closed?
- How do new technologies and public use of these technologies affect network traffic?
- What is the future role of sensors within an alerting system?
- WEA is currently limited to text only. What additional technologies are needed to extend WEA to include either images or links to maps?
- As recipients receive messages on their various computing and mobile devices, they may wish to forward those messages to others via text or email or to social media sites. Additionally, they may want to simply link to additional information. How can these capabilities be incorporated in alerting systems? What strategies and techniques can be used to decrease bandwidth requirements?
- How can existing systems and new technologies incorporate the needs of at-risk populations, including those with physical or mental disabilities?

RESPECTING PRIVACY AND MEETING SECURITY NEEDS

Concerns over privacy are prevalent in discussions surrounding the use of mobile devices and the use of geolocation information.

- Users make privacy trade-offs frequently. They enjoy the use of a particular mobile application that may share their location information with others, often without explicit consent. If more explicit consent is needed for applications to share information, do users have the knowledge to make this decision?
- How can alerting systems be designed up front to incorporate privacy and security concerns (versus trying to incorporate privacy and safety considerations after the system is designed, or worse, after it is deployed)?
- What is the likelihood that concerns about negative public relations or apprehensiveness regarding government regulation will discourage developers from incorporating geolocation information into emerging technologies?

FACILITATING AND ENCOURAGING USE BY PRACTITIONERS

Ultimately, practitioners at various levels of government decide how and when to send an alert. Clear guidelines may be required to encourage more and appropriate use of new systems. Additional questions include the following:

- What are possible incentives for emergency managers to experiment with the use of new alerting systems and geotargeted alerts and warnings? What are the major constraints that limit adoption by local practitioners?
- What policy framework would help encourage the use of new technologies for alerting by practitioners?
- What are useful ways to involve more practitioners in technology and system design and decision-making processes?

Appendixes

A

Workshop Agenda

FEBRUARY 21-22, 2013
NATIONAL ACADEMIES KECK CENTER
WASHINGTON, D.C.

Day 1: Public Response and Considerations for Geotargeted Alerts and Warnings

Past research has shown that specific and clear information, including which locations are and are not at risk, increases the likelihood that people take protective action. When alerts and warnings are delivered to broader populations than those actually affected by an event, the result may be that an alert or warning indicating more people than are actually at risk should take action. With new technological opportunities to more precisely target alerts and warnings come new questions about public response:

- What degree of geographical targeting is needed to make messages relevant? In what scenarios might greater precision be useful?
- What is known about the consequences of too many messages (e.g., if the threshold for events which trigger alerts is set too low, if alerts cover too large a geographical area, if messages are repeated too often, or if there are too many false alarms)? Is there a threshold above which people will ignore messages or opt out from receiving them?
- What are potential drawbacks of better geotargeting capabilities, such potential for privacy protections?

8:30 am	**Welcome**

> Ellis Stanley, Chair, Committee on Geotargeted Disaster Alerts and Warnings
> Dan Cotter, Geospatial Information Officer, Department of Homeland Security
> Denis Gusty, Science and Technology Directorate, Department of Homeland Security

9:00	**Overview of Past CSTB Alerts and Warning Work**

Public Response to Alerts and Warnings on Mobile Devices
Jeannette Sutton, Chair, Committee on Public Response to Alerts and Warnings on Mobile Devices

Public Response to Alerts and Warnings Using Social Media
Leslie Luke, Committee on Public Response to Alerts and Warnings Using Social Media

9:45	**Value of Geotargeted Alerts and Warnings**

Moderator: Dennis Mileti

What Role Does Geotargeted Information Play in Effectively Communicating Risks to At-Risk and Not-At-Risk Populations?
Tim Sellnow, University of Kentucky

What Are the Various Ways that Geotargeted Information Can Be Communicated to the Public? Under What Circumstances Might One Method Be Preferred Over Another?
Michele Wood, California State University, Fullerton

For What Hazards and Protective Actions Is Geotargeting Most Needed?
Brooke Liu, University of Maryland

How Do Present-Day Tools Constrain Emergency Managers? Are Some Deployed Capabilities Being Underused?
Ken Rudnicki, City of Fairfax, Virginia

11:45	**Geotargeting Needs and Challenges for Particular Hazards**

Moderator: Ellis Stanley

Wildfire Events
Thomas Cova, University of Utah

Radiological/Nuclear Incident
Steven M. Becker, Old Dominion University College of Health Sciences

Transportation Systems
Peter LaPorte, Washington Metropolitan Area Transit Authority

12:45 pm	Lunch
2:00	**Data Security and Privacy**

Moderator: Ming-Hsiang Tsou

Mobile Device Privacy and Security Concerns
Patrick McDaniel, Pennsylvania State University

Personal Privacy
Marc Armstrong, University of Iowa

Methods for Preserving Privacy While Providing Geotargeted Alerting
Darrell Ernst, Private Consultant

Legal Questions Surrounding Location Information
Kevin Pomfret, Centre for Spatial Law and Policy

3:30	**Location-Enabled Technologies—Part 1**

Moderator: Shashi Shekhar

Wireless Location Determination
Larry Dodds, TruePosition

Indoor Position Technologies
Ayman Naguib, Qualcomm

4:30 **Day 1 Summary and Discussion**

Ellis Stanley, Chair, Committee on Geotargeted Disaster Alerts and Warnings
Dennis Mileti, University of Colorado, Boulder; Committee Member

Day 2: Technologies and Tools for More Precise Geotargeted Alerts and Warnings

Cell phones and other mobile devices can determine their position using cell tower triangulation, GPS, and nearby Wi-Fi sites and offer ample computing power and high-resolution displays to receive, process, and display alerts and warnings. Similarly, other computing devices such as laptops, desktops, and cable set top boxes can also establish their location and with suitable software provide targeted alerts.

- How can already-deployed and emerging technologies be used to deliver improved geographical targeting capabilities?
- What would be effective strategies for introducing more precise geographic information as systems are modernized and enhanced?
- What technical and operational standards are needed to facilitate the delivery of more precise alerts/warnings?
- How can commercial off-the-shelf technology and commercial services be leveraged to deliver alerts and warnings?

8:30 am **Current and Future Vision for the Integrated Public Alert and Warning System**
Moderator: Art Botterell

Mike Gerber, National Weather Service
Denis Gusty, S&T Directorate, Department of Homeland Security
Wade Witmer, IPAWS Division, Federal Emergency Management Agency

9:15 **Lessons from and Opportunities for Traditional Technologies for Geotargeted Alerts**
Moderator: Helena Mitchell

Telephone Alerting
Rick Wimberly, Galain Solutions

APPENDIX A

Radio Broadcast Technologies
John Kean, NPR Labs

Weather Radio Technologies
Bruce Thomas, Midland Radios (remotely)

Cable Television Alerting
Ron Boyer, Boyer Broadband

10:30 **Location-Enabled Technologies—Part 2**
Moderator: Mani Chandy

Geotargeted Alerts and Warnings in Streaming Video
Hisham Kassab, MobiLaps

Geotargeting with Internet Protocols
Richard Barnes, BBN Technologies/Raytheon, IETF Geographic Location Working Group

11:45 **Current and Future Capabilities of Location-Enabled Mobile Devices**
Moderator: Ramesh Rao

Geotargeting of SMS
George Percivall, Open Geospatial Consortium

Carrier Capabilities
John Davis, Sprint

Third-Party Application Capabilities
J.T. Johnson, Weather Decision Technologies

Mobile Location Determination
Farshid Alizadeh, Skyhook Wireless (remotely)

1:00 pm **Wrap-Up Discussion**

Ellis Stanley, Committee on Geotargeted Disaster Alerts and Warnings
Denis Gusty, Department of Homeland Security

B

Biosketches of Workshop Speakers

Farshid Alizadeh-Shabdiz, chief scientist for Skyhook, is responsible for the research and development of Skyhook's positioning technology. Dr. Alizadeh-Shabdiz has almost 20 years of industrial experience in the design and implementation of satellite and wireless networks. Before joining Skyhook, he was the head of the communications section of Advanced Solutions Group (part of Cross Country Automotive Services). There, he was responsible for the management, design, and implementation of an application server and media gateway. Dr. Alizadeh-Shabdiz was a member of the design and implementation team of the first three satellite-based mobile networks at Hughes Network Systems: ICO, Thuraya, and Inmarsat high-speed data network. He proposed the first complete analytical model to carry out an analysis of single-hop and multihop ad hoc networks and 802.11 based on WLANs. Dr. Alizadeh-Shabdiz is on the faculty of Boston University and received his Ph.D. from George Washington University and his M.Sc. from Tehran University.

Marc P. Armstrong is a professor in the Department of Geography at University of Iowa. During the 2012-2013 academic year, he served as interim chair of the Department of Communication Studies and as interim chair of the Department of Cinema and Comparative Literature. Dr. Armstrong also holds a courtesy appointment in the Graduate Program in Applied Mathematical and Computational Sciences. He was named a College of Liberal Arts and Sciences (CLAS) collegiate fellow in 2005 and he served as interim associate dean for research in CLAS in 2006, as interim director of Iowa's School of Journalism and Mass Communication in 2007 and 2008, and as interim director of the Division of World Languages, Literatures, and Culture in 2010-2011. Dr. Armstrong's Ph.D. is from the University of Illinois, Urbana-Champaign. A primary focus of his research is on the use of cyberinfrastructure to improve the performance of spatial analysis methods. Other active areas of interest focus on the use of geospatial technologies by groups and geographic aspects of privacy. Dr. Armstrong has served as North American editor of the *International Journal of Geographical Information Science*, served on the editorial boards of six journals, and has published more than 100 academic papers, including articles in a wide variety of peer-reviewed journals such as *Annals of the Association of American Geographers, Photogrammetic Engineering and Remote Sensing, Geographical Analysis, Statistics and Medicine, Mathematical Geology, Computers and Geosciences, International Journal of Geographical Information Science, Parallel Computing, Computers, Environment and Urban Systems*, and *Journal of the American Society for Information Science*.

Richard Barnes is a researcher with BBN Technologies. He leads BBN's Internet standards efforts in the areas of geolocation, presence, and emergency services. He is chair of the IETF GEOPRIV working group, a former chair of the ECRIT working group, and was recently appointed to be the IETF area director for real-time applications and infrastructure.

Steven M. Becker is professor of community and environmental health in the College of Health Sciences at Old Dominion University. He is a leading international expert on community responses to unconventional disasters, public health preparedness and response, and risk communication and emergency messaging for chemical, biological, radiological, and nuclear issues. Dr. Becker served as a principal investigator (PI) in the Centers for Disease Control and Prevention–Association of Schools of Public Health Pre-Event Message Development Project, one of the most extensive peer-reviewed studies ever conducted of people's concerns and communication needs in situations involving unconventional health threats. More recently, he has served as PI for a multiyear Department of Homeland Security (DHS) study of the communication and information challenges posed by radiological threats and incidents. In addition to his scholarly research, Dr. Becker has extensive field experience at the sites of major incidents around the world, including such cases as a major drinking-water contamination incident in Great Britain; the 1999 nuclear criticality accident in Tokaimura, Japan; and the 2001 foot-and-mouth disease outbreak in the United Kingdom. He has also done follow-up work in Ukraine and Belarus on the community impacts of the Chernobyl disaster. In 2011, Dr. Becker was a member of a three-person radiological emergency assistance team invited to Japan in response to the earthquake-tsunami disaster and the accident at Fukushima Dai-ichi nuclear plant. While on the ground, the team carried out a rapid site assessment in affected areas, exchanged information with Japanese disaster response organizations, and provided training to more than 1,100 Japanese physicians, nurses, and other healthcare providers and emergency responders. In 2005, Dr. Becker was elected to serve on the National Council on Radiation Protection and Measurements, and his work on emergency management and risk communication has been recognized by such scientific organizations as the Health Physics Society and Oak Ridge Associated Universities. He has also been a visiting fellow at the Japan Emergency Medicine Foundation and National Hospital Tokyo Disaster Medical Center. For more than a decade, Dr. Becker has been an invited faculty member for Harvard School of Public Health's course on radiological emergency planning. Early in 2012, he was named to the Thought Leader Advisory Council of the National Public Health Information Coalition. In September 2012, Dr. Becker was appointed by President Barack Obama to the U.S. Nuclear Waste Technical Review Board.

Ron Boyer serves the cable telecommunications industry, having spent more than 35 years as an engineer, working in both the manufacturing and operations sides. He has experience in virtually all engineering aspects that the industry has to offer. Prior to starting his consulting firm Boyer Broadband in 2011, Mr. Boyer worked for Time Warner Cable, the second-largest cable operator in the United States. He held various positions in their corporate management. At first as a senior staff engineer, then as senior network engineer, and finally more than 7 years as senior regulatory engineer in the legal department. This experience provided him a solid working understanding of both the technology utilized and regulatory environments involved in the day-to-day operations of a cable system. Before joining Time Warner Cable in 1997, Mr. Thomas also worked for ADC Broadband and Scientific Atlanta (now a division of Cisco) for more than 11 years. This experience proved invaluable when working for Time Warner Cable; it provided a firm understanding of how important the communications between the manufacturer and the user are when deploying advanced technologies. He has maintained an association with a diverse range of organizations, including the Society of Cable Telecommunications Engineers, California Public Utilities Commission, IEEE National Electrical Safety Code and National Electrical Code committees, and Communications Security, Reliability and Interoperability Council (CSRIC). He has participated as a member or held various lead positions in the different working groups. Recent activities included participating in two CSRIC III Working Groups (2 and 9).

Dan Cotter is the director of the Information Applications Division of the First Responder Group (FRG) in the Science and Technology Directorate, DHS. Mr. Cotter is also the DHS geospatial information officer and senior agency official for geospatial information. FRG identifies, validates, and facilitates the fulfillment of first responder capability gaps through the use of existing and emerging technologies, knowledge products, and the acceleration of standards. FRG engages first responder working groups, teams, and other stakeholders to better understand the needs and requirements of the first responder communities. Prior to joining the FRG, Mr. Cotter served as the DHS chief technology officer (CTO). As the DHS CTO, his responsibilities included overseeing programs for information sharing, enterprise architecture, enterprise data management, geospatial technologies, identity, credentialing and access management, as well as the Homeland Security Information Network (HSIN) and the DHS Common Operational Picture investments. Mr. Cotter served as the DHS geospatial management officer from 2005 to 2007. In fall 2005 he was deployed to the Katrina-Rita Joint Field Office, Baton Rouge, Louisiana, to serve as the Geospatial Intelligence Unit manager. Mr. Cotter's private sector experience includes acting as the geospatial information technologies manger for a large engineering firm, as the president of an airborne light detection and ranging (lidar) company, and as vice president of a flood zone determination firm. His prior public sector experience includes 12 years with the Federal Emergency Management Agency (FEMA) applying geospatial and remote sensing technology to natural hazard mitigation programs, including the National Flood Insurance Program, and disaster response, including Hurricane Andrew, the Northridge Earthquake, and the 1993 Midwest Floods. Mr. Cotter was elected as a fellow of the American Association for the Advancement of Science in 2005. He has received numerous awards, including the FEMA Director's Distinguished Service Award, the National States Geographic Information Council Outstanding Service Award, the Transamerica Pyramid Award for Business Reengineering, and the NASA National Resources Award for lidar commercialization. Mr. Cotter holds an M.B.A. from Texas A&M University, an M.S. in geographic and cartographic sciences from George Mason University, a B.S. in hydrology from the University of Arizona, an A.A.S. in computer information systems from Northern Virginia Community College, and a Federal Chief Information Officer Graduate Certificate from the University of Maryland, University College.

Thomas Cova is professor of geography and director of the Center for Natural and Technological Hazards at the University of Utah, Salt Lake City. He has a B.S. in computer science from the University of Oregon and an M.A. and Ph.D. in geography from the University of California, Santa Barbara. His research and teaching interests are environmental hazards, transportation, and geographic information science with a particular focus on wildfire evacuation modeling, analysis, and planning. He has published on a variety of topics in many leading hazards, transportation, and geographic information system (GIS) science journals and is most known for work on evacuation vulnerability and routing in fire-prone communities of the western United States. He has served as chair of the GIS Specialty Group of the Association of American Geographers (AAG) and as program chair for the International Conference of Geographic Information Science (GIScience '08) and currently chairs the AAG Hazards, Risks, and Disasters Specialty Group. He teaches courses on hazards geography, emergency management, and GIS.

John Davis is the lead network design and development engineer at Sprint responsible for implementation of the Commercial Mobile Alert Service (CMAS) platform within the Carrier network. Sprint was an early adopter of CMAS, and Mr. Davis led the first large-scale test at the Carrier level with the County of San Diego. He has been with Sprint for 12 years helping to develop the wireless Internet since its genesis in addition to leading all CMAS initiatives. He is currently Sprint's representative on the CSRIC III Working Group 2 committee.

Larry Dodds is currently the vice president of product line management and business development at TruePosition. In his current role, Mr. Dodds is responsible for all of TruePosition's location solutions and forging new partner relationships. He has more than 20 year of experience in location, including star-based navigation for the U.S. Navy's sea-launched ballistic missiles, the initial test program for the U.S. Air Force GPS receivers, and now various forms of location techniques for mobile devices. He has a bachelor's degree in electrical engineering from Drexel University and a master's degree in computer science from Northeastern University.

Darrell Ernst is an advisor to the office in the Pentagon responsible for the development of the instrumentation systems used on U.S. test ranges for the testing of weapon systems. He advises the deputy director of the office on technical and regulatory issues on radio spectrum used at the test ranges. He advises on the implementation of C-band for telemetry and range spectrum encroachment. He led planning for investment in research and development of technologies for more efficient use of the radio spectrum. He was a member of the U.S. delegation to the 2007 and 2012 World Radiocommunication Conferences. He began his career in 1961 as a telemetry technician in tracking aircraft operating on the Air Force Eastern Test Range. Mr. Ernst was involved in the test programs for Atlas, Titan, Polaris, Poseidon, Projects Mercury, Gemini, Apollo, the space shuttle, the Global Positioning System (GPS) user segment, and many other space programs. In 1998 his employer, the MITRE Corporation, asked him to work with the emergency management community to explore the possibility of using the technology he and others had invented for public warning. Working with various leaders in the emergency management community, he came up with the idea of convening a workshop to identify the requirements for a modern emergency warning system. He was at the 2001 annual meeting of the National Emergency Management Association meeting in Montana on September 11 to announce the workshop, which was scheduled for that November. As a consequence of the events of that day, the workshop became a national referendum that resulted in the creation of the Partnership for Public Warning (PPW). MITRE assigned him to manage the PPW. One of the achievements of PPW was its sponsorship of Art Botterell's Common Alerting Protocol (CAP) standard project that included an important national workshop on CAP at the National Emergency Training Center in Emmitsburg, Maryland. Mr. Ernst graduated from Auburn University with a bachelor of science degree in mathematics and graduated from Rensselaer Polytechnic Institute with a master's degree in operations research and statistics. He retired from the U.S. Air Force in 1982. Mr. Ernst is a co-inventor of a concept for message distribution using spatial coordinates for addressing, thereby obviating the need to know the address or location of message recipients. The concept has been embodied in various prototypes such as the Tactical Automated Situation Receiver for the U.S. Army and a cell-phone-based alerting system for a major telecommunications company. He is listed as an inventor on three patents issued to MITRE.

Mike Gerber is the emerging dissemination technologies lead for the National Oceanic and Atmospheric Administration's (NOAA's) National Weather Service (NWS) in Silver Spring, Maryland. He joined the NWS in 1992. Mr. Gerber is leading efforts to bring about the integration of NWS alert information across the widest possible range of warning systems and consumer electronic devices to better save lives, protect property, and enhance the national economy. Mr. Gerber brings a visionary perspective as the NWS representative on several cross-organizational teams working to improve the Wireless Emergency Alert (WEA) system. He also leads efforts within the NWS to improve mobile alerting through enhancements to NWS alert generation tools and CAP. Mr. Gerber is a senior fellow of the Council for Excellence in Government. He made significant contributions to the land and fire management community as a former fire weather and incident meteorologist at the NWS Forecast Office in Boise, Idaho. Mr. Gerber spearheaded development of weather forecast guidance for hundreds of weather observation stations that improved prescribed fire planning and wildfire prediction throughout the western United States. While working as a meteorologist at the NWS Forecast Office in Sterling, Virginia, Mr. Gerber co-anchored the PBS television show, *AM Weather*. Mr. Gerber earned a bachelor's degree in atmospheric science from the University of Arizona.

Denis Gusty serves as the program manager for the FRG's Commercial Mobile Alert Service (CMAS) Research, Development, Testing, and Evaluation (RDT&E) Program. CMAS RDT&E is responsible for improving CMAS's capabilities, including geotargeting and how the public responds to wireless emergency alerts. In addition, Mr. Gusty leads FRG's Emergency Data Exchange Language Program, which focuses on improving messaging standards that help emergency responders manage incidents and exchange information in real time. Mr. Gusty came to DHS Science and Technology Directorate from the U.S. General Services Administration (GSA), where he served as director of GSA's Office of Intergovernmental Solutions. Prior to joining GSA, Mr. Gusty served as a program manager at the U.S. Department of Labor. In this role, he was responsible for helping to implement the President's Management Agenda by managing the e-government initiative, GovBenefits.gov. Mr. Gusty has 15 years of experience in developing intergovernmental partnerships and information technology policy and practices.

J.T. Johnson co-founded Weather Decision Technologies, a global weather information company helping businesses and individuals make decisions related to weather, in 2000 where he currently serves as the chief technology officer. Before joining Weather Decision Technologies, Mr. Johnson was a team leader at NOAA's National Severe Storms Laboratory and a science and operations officer at the Olympic Weather Support Office. He completed his B.S. and M.S. degrees in meteorology at the University of Oklahoma.

Hisham Kassab is the founder and president of MobiLaps, LLC, a high-tech company developing innovative technologies to power/enable/enhance next-generation alert dissemination channels, with a current emphasis on broadband alerts (including streaming media), CMAS/WEA/Personal Localized Alerting Network, and social media alerting.

Dr. Kassab has 20 years of experience in the information and communications technology (ICT) industry, with the last 5 focused on next-generation alerting. He has been an active member of an advisory group to the Federal Communications Commission on next-generation alerting (CSRIC III-Working Group 2). Prior to MobiLaps, Dr. Kassab worked as a strategy and technology consultant with Booz Allen Hamilton (now Booz and Co.) focusing exclusively on ICT clients. He earned his B.S., M.S., and Ph.D. degrees in electrical engineering and computer science from the Massachusetts Institute of Technology (MIT). He also holds an M.S. in operations research from MIT. His doctoral dissertation was in the area of wireless data networks.

John Kean, a senior technologist for NPR Labs, develops and supervises the technical projects of NPR Labs, the only not-for-profit broadcast engineering laboratory in the United States, which is involved in the development and evaluation of new technologies, procedures, and standards on behalf of public radio. Mr. Kean was a senior engineer at NPR from 1980 to 1986, where he supported new broadcast technologies and pioneered expansion of FM subcarrier services. He left NPR to join Jules Cohen and Associates and get his start in consulting engineering. From 1987 to 2000 he was a director of engineering for Moffet Larson and Johnson, Inc., consulting in the fields of broadband wireless networks, TV and radio facilities, FCC regulations, and microwave and satellite systems. Before returning to NPR in 2004, Mr. Kean was director of wireless architecture for XO Communications, a broadband telecommunications company having extensive broadband wireless holdings. He is a member of IEEE and past president of the IEEE Broadcast Symposium, contributing author to *The NAB Engineering Handbook,* Editions 7, 8, and 9, and presenter of numerous papers in the field of radio systems engineering to the National Association of Broadcasters' Engineering Conference, International Engineering Consortium, Wireless Communications Association, and has served as a delegate to the International Telecommunication Union plenary meetings in Geneva on behalf of the North American Broadcasters Association. He is past president of the Audio Engineering Society (Washington DC Section), co-chair of the National Radio Systems Committee's AM Study Task Group, a recent member of the Consumer Electronics Association's Audio Division Board, and has a patent pending for the prediction of coverage for U.S. in-band on-channel digital audio broadcasting. His recent work has focused on digital audio broadcasting, including digital audio codec performance, HD Radio® multicast developments, overall broadcast system performance, and the prediction of broadcast signal transmission and reception.

Peter LaPorte currently serves as the director of the Office of Emergency Management (OEM) of the Washington Metropolitan Area Transit Authority (WMATA). WMATA established its OEM within the Metro Transit Police Department in 2009 to institute an emergency management mindset and culture, and the OEM team has seen emergency management awareness and practices become more common at WMATA in the past 3 years from the development of an emergency operations plan (EOP), continuity of operations plans, a terrorism incident annex to the EOP, rail station emergency response plans, and more.

Brooke Liu is an associate of communication at the University of Maryland and a research affiliate with the National Consortium for the Study of Terrorism and Responses to Terrorism (START). Dr. Liu's research primarily examines how governments manage communication during crisis and noncrisis situations. Her research has been published in outlets such the *Handbook of Crisis Communication, Journal of Applied Communication Research, Journal of Communication Management, Journal of Public Relations Research,* and *Natural Hazards Review*. She received her Ph.D. in mass communication from the University of North Carolina, Chapel Hill (2006), and M.A. in journalism from the University of Missouri, Columbia (2003). In recent years, Dr. Liu has served as a public affairs volunteer for the American Red Cross for the Arlington, Virginia, and Chicago, Illinois, chapters as well as a research consultant for national headquarters. As part of her work at START, she currently leads four DHS-funded projects focusing on effective risk communication and messaging. She also continues to provide research support as an independent consultant primarily to government agencies, most recently focusing on evaluating social media campaigns.

Leslie Luke is the group program manager for the County of San Diego's Office of Emergency Services, where he oversees the Planning Branch, Info/Intel Branch, Recovery Branch, and Support Services. Mr. Luke is the recovery coordinator for the County of San Diego and has been the recovery operational area lead for five federally declared disasters and numerous state-declared disasters. He coordinates the Continuity of Community Programs and is a liaison with schools, including child care resource centers, the business sector (leads the ReadySanDiego Business Alliance), and faith-based initiatives. He oversees the office's public awareness/public education initiatives, special projects, and the student worker/internship/volunteer program. Mr. Luke has worked for the County of San Diego for 22 years, in the Office of Emergency Services since 2004. Prior to that, he worked in the Public Safety Group, a division of the

County's Chief Administrative Office, and was an investigator for the County Medical Examiner's Office.

Patrick McDaniel is a professor in the Computer Science and Engineering Department at the Pennsylvania State University and co-director of the Systems and Internet Infrastructure Security Laboratory. Dr. McDaniel's research efforts centrally focus on network, telecommunications, and systems security, language-based security, and technical public policy. He is the editor-in-chief of the Association for Computing Machinery (ACM) journal *Transactions on Internet Technology* and serves as associate editor of *Transactions on Information and System Security* and *IEEE Transactions on Computers*, and he stepped down as associate editor of *IEEE Transactions on Software Engineering* in 2012. Dr. McDaniel was awarded the National Science Foundation (NSF) CAREER Award and has chaired several top conferences in security including, among others, the 2007 and 2008 Institute of Electrical and Electronics Engineers (IEEE) Symposium on Security and Privacy and the 2005 USENIX Security Symposium. Prior to pursuing his Ph.D. in 1996 at the University of Michigan, he was a software architect and project manager in the telecommunications industry.

Ayman Naguib received a B.Sc. degree (with honors) and a M.S.EE degree in electrical engineering from Cairo University, Cairo, Egypt, in 1987 and 1990, respectively, and an M.S. degree in statistics and Ph.D. degree in electrical engineering from Stanford University in 1993 and 1996, respectively. From 1996 to 2000, Dr. Naguib was a principal member of technical staff at AT&T Shannon Labs, where he, along with his colleagues at AT&T Labs, pioneered the field space-time coding. From September 2000 to August 2002, he was with Morphics Technology, Inc. In October 2002, Dr. Naguib joined Qualcomm, Inc., where he is now a director of engineering with Qualcomm Research, Silicon Valley, where he is currently leading indoor positioning and navigation research activities. His 1998 *IEEE Journal on Selected Areas in Communications* paper on space-time coding was selected by the IEEE Communication Society as one of the 50 fundamental papers ever published by the society. His 2003 *IEEE JSAC* paper won the best paper award. He has 40 U.S. patents, more than 90 pending patent applications, and more than 50 book chapter, conference, and journal publications. Dr. Naguib served as an associate editor for *IEEE Transactions on Communications* from 2002 to 2007 and as a guest editor to a number of IEEE transactions journals. In 2006, Dr. Naguib was named an IEEE fellow for his contributions to space-time coding and signal processing and wireless communications. His current research interests are statistical learning, location determination, and indoor positioning.

George Percivall is an accomplished leader in geospatial information systems and standards. As chief engineer of the Open Geospatial Consortium (OGC), he is responsible for the OGC Interoperability Program and the OGC Compliance Program. His roles include articulating OGC standards as a coherent architecture, as well as addressing implications of technology and market trends on the OGC baseline. Prior to joining OGC, Mr. Percivall was chief engineer with Hughes Aircraft for NASA's Earth Observing System Data and Information System—Landsat/Terra release; principal engineer for NASA's Digital Earth Office; and he represented NASA in OGC, International Organization for Standardization, and Committee on Earth Observation Satellites. He was director of the Global Science and Technology's Geospatial Interoperability Group. Previously, he led developments in intelligent transportation systems with the U.S. Automated Highway Consortium and General Motors Systems Engineering, including the EV1 program. He began his career with Hughes as a control system engineer on GOES/GMS satellites. He holds a B.S. in engineering physics and an M.S. in electrical engineering from the University of Illinois, Urbana-Champaign.

Kevin Pomfret is the executive director of the Centre for Spatial Law and Policy and the founder of GeoLaw, P.C. He is well known within the spatial technology community for his efforts to increase the dialogue on the legal and policy issues associated with spatial data. Mr. Pomfret has worked with and around spatial technology for more than 20 years. Prior to attending law school, Mr. Pomfret served as a satellite imagery analyst with the U.S. government. In that capacity, he developed an imagery collection strategy to monitor critical arms control agreements and worked on requirements for future collection systems. In addition, he served as the special assistant to the U.S. government official responsible for developing the intelligence community's satellite imagery collection and exploitation requirements. Upon entering private practice, Mr. Pomfret recognized that there were a number of unique legal issues associated with spatial data, including intellectual property rights, licensing, liability, privacy, and national security. He regularly advises a variety of spatial technology companies on such matters as licensing and distribution agreements, privacy policies, and spatial data audits. He also works as a consultant on developing a legal and policy framework for national spatial data infrastructures.

Ken Rudnicki has more than 35 years of experience in emergency management. Mr. Rudnicki began his career in emergency management in 1977 while a member of the U.S. Air Force. After completing the Air Force Disaster Preparedness School, he was stationed in Fort Walton Beach,

Florida, at Hurlburt Field where his office was awarded Best Disaster Preparedness Program in Tactical Air Command. Throughout his Air Force career he was stationed across the globe and has responded to a wide variety of disasters, including earthquakes in California, typhoons in the far east, and volcanoes in the Philippines. He was awarded two Humanitarian Service Medals, five Air Force Commendation Medals, and a Meritorious Service Medal during his 24-year career in disaster preparedness. Following his retirement from the Air Force, Mr. Rudnicki joined the Florida Division of Emergency Management where he worked as a planner for 4 years and the area coordinator for the Tampa Bay region for 6 years. During this time, he was involved in more than 20 federal disaster declarations. While working for Florida, he was awarded three Distinguished Service Awards. As a member of the Florida Emergency Preparedness Association, he helped develop regional response team procedures bringing local emergency managers to the assistance of impacted counties and the state. Mr. Rudnicki then moved into the private sector as a consultant working in Reston, Virginia, where he developed numerous domestic security exercises, developed state and local plans, and was selected by the Secretary of DHS to be part of the Nationwide Plan Review ordered by the President following Hurricane Katrina. In 2006, Mr. Rudnicki accepted a job with the City of Fairfax, Virginia, as the emergency coordinator and has remained in this position to date. He has served on numerous committees throughout his career and now serves as the president of the Virginia Emergency Managers Association, and he is the International Association of Emergency Managers (IAEM) USA-Region III secretary and treasurer. Mr. Rudnicki is a certified professional emergency manager with the states of Florida and Virginia and a certified emergency manager through the IAEM.

Timothy L. Sellnow is a professor of communication at the University of Kentucky, where he teaches courses in risk and crisis communication. Dr. Sellnow's research focuses on bioterrorism, pre-crisis planning, and communication strategies for crisis management and mitigation. He has conducted funded research for DHS, the U.S. Department of Agriculture, and the Centers for Disease Control and Prevention. He has published numerous refereed journal articles on risk and crisis communication and has co-authored four books on risk and crisis communication. His most recent book is *Risk Communication: A Message-Centered Approach*. He is also past editor of the National Communication Association's *Journal of Applied Communication Research*. Dr. Sellnow received his Ph.D. from Wayne State University in 1987.

Jeannette Sutton is a senior research scientist in the Trauma Health and Hazards Center at the University of Colorado, Colorado Springs, where she specializes in disaster sociology with a primary focus on online informal communications in disaster, public alerts and warnings, and community resiliency. Much of her research investigates the evolving role of information and community technology, including social media and mobile devices, for disaster preparedness, response, and recovery. Dr. Sutton is the principal investigator (PI) on two NSF-funded projects, one on the use of Twitter for disaster communications and a second on the role of information access in relation to perceptions of collective efficacy. She is also a co-investigator on the DHS-sponsored project Comprehensive Testing of Imminent Threat Public Messages for Mobile Devices. Dr. Sutton holds a Ph.D. in sociology from the University of Colorado, Boulder, and completed her postdoctoral training at the Natural Hazards Center. She is also a special term appointee with the Center for Integrated Emergency Preparedness at Argonne National Laboratory.

Bruce Thomas has served as chief meteorologist and national spokesperson for Midland Radio Corporation, Kansas City, Missouri, since 2004. Mr. Thomas has been recognized by the Department of Commerce with the Mark Trail Award for outstanding service promoting *All Hazards NOAA Weather Radio* across America. Prior to his work with Midland Radio, Mr. Thomas spent nearly two decades as a broadcast meteorologist in Tornado Alley, working with network affiliate television stations in College Station, Waco, Dallas/Fort Worth, and Kansas City. He is currently serving as president of the National Weather Association. He is also an active member of the American Meteorological Society where he holds the designation Certified Broadcast Meteorologist.

Rick Wimberly is president of Galain Solutions, Inc., an independent consultancy with expertise in alerts and warnings, which serves clients at local, state, and federal levels as well as private industry. Galain's clients include the FEMA Integrated Public Alert and Warning System (IPAWS) program. Mr. Wimberly has been involved in the alert and warning industry for 15 years and in the public safety industry for nearly 25 years. He writes extensively on topics related to alerts and warnings, including a widely followed blog for *Emergency Management* magazine. A recent cover story article Mr. Wimberly wrote for *Emergency Management* magazine, "Do Alert Notifications Fail to Live Up to Expectations," addressed common shortcomings of telephone-based alerting systems, including challenges with geotargeted messages.

Wade Witmer has been with the FEMA IPAWS Division since January 2009. The IPAWS program is tasked with implementing the vision of Executive Order 13407 for the United States to have "an effective, reliable, integrated, flexible, and comprehensive system to alert and warn the American people." The IPAWS brings together the Emergency Alert System, the new CMAS, a feed for publishing alerts to Internet services, and integration with NWS's All-Hazards Radio network. Using industry standard protocols, authorized public safety officials can use IPAWS to send emergency alerts to citizens in their local area. Prior to joining the IPAWS Division at FEMA, Mr. Witmer was employed with the Defense Information Systems Agency for 9 years, serving across various programs as a communications systems engineer, program manager, and portfolio manager for Mobile Communications in the Presidential Communications Upgrade Program. Just prior to joining FEMA, he served as the White House Communications Agency deputy director of enterprise architecture, strategic planning, and systems engineering. Mr. Witmer has more than 20 years of experience in government systems engineering and program acquisition management. He has a bachelor of science degree in electrical engineering from the Pennsylvania State University.

Michele Wood is an assistant professor in the Health Science Department at the California State University, Fullerton, where she teaches courses in statistics and program design and evaluation. Dr. Wood has 20 years of experience designing, implementing, and evaluating interventions. Through her affiliation with the Southern California Injury Prevention Center in the University of California, Los Angeles (UCLA) School of Public Health, she managed a national household preparedness survey conducted as part of the National Center for the START program through the University of Maryland's Center of Excellence, as well as a California household telephone survey of earthquake preparedness. Dr. Wood received her Ph.D. in public health from the Department of Community Health Sciences at UCLA, and she also holds a master's degree in community psychology.

C

Biosketches of Committee Members

Ellis M. Stanley, Sr., *Chair,* is the former vice president, emergency management, disaster and mitigation at Dewberry, LLC, and has more than 32 years of work experience in emergency management beginning as director of emergency management for Brunswick County, North Carolina, in 1975. Mr. Stanley was selected as the first fire marshal for Brunswick County and served as fire and rescue commissioner and was very involved with hurricane planning and response as well as having developed one of the first fixed nuclear facility plans in the United States following Three Mile Island. Mr. Stanley was appointed in 1982 as the director of the Durham-Durham County Emergency Management Agency where he worked very closely with the world's largest research park in the North Carolina Triangle area and was heavily involved with hazardous materials planning. In 1987 Mr. Stanley was appointed by the Governor of Georgia as the director of the Atlanta-Fulton County Emergency Management Agency. While in Atlanta, Mr. Stanley had extensive experience in major event planning (1988 Democratic National Convention (DNC), 1995 Mandela visit, and the 2006 International Olympic Games). He was appointed in 1997 as assistant city administrative officer for the City of Los Angeles and then in 2000 as general manager of the Emergency Preparedness Department for the City of Los Angeles until his retirement in 2007. Mr. Stanley joined Dewberry, LLC, in November 2007 as director of Western Emergency Management Services. In March 2008, he was selected to be the director of DNC planning for the City and County of Denver, Colorado. He received his B.S. in political science from the University of North Carolina, Chapel Hill.

Art Botterell is a research scientist at Carnegie Mellon University's Silicon Valley campus. His experience in emergency public information and public warning spans more than four decades, including service with the Federal Emergency Management Agency, the California Emergency Management Agency, and local public safety and emergency management agencies. Mr. Botterell has served as a consultant to the Department of Homeland Security and other federal agencies, as well as in North America, Europe, Australia, and Asia, and with the United Nations Development Programme. He served as a member of the Federal Communications Commission's (FCC's) Commercial Mobile Alerting Advisory Committee and a variety of other government and scientific panels, including an National Research Council committee. Mr. Botterell was a founding trustee of the nonprofit Partnership for Public Warning. He has also worked as a broadcast engineer, a journalist, and an online content producer. He originated and guided the development of the Common Alerting Protocol standard.

K. Mani Chandy (NAE) is the Simon Ramo Professor at the California Institute of Technology (Caltech). Dr. Chandy has worked for Honeywell and IBM. From 1970 to 1989, he was in the Computer Science Department of the University of Texas, Austin, serving as chair in 1978-1979 and 1983-1985. He has served as a consultant to a number of companies, including IBM and AT&T Bell Labs. He has been at Caltech since 1987, 2 years as a Sherman Fairchild Fellow and then as a professor in computer science. Dr. Chandy is a member of the National Academy of Engineering. He received the IEEE Koji Kobayashi Award for Computers and Communication in 1987, the A.A. Michelson Award from the Computer Measurement Group in 1985, and has numerous other awards. Software developed by Dr. Chandy and colleagues in the area of computer performance modeling was marketed by Boole and Babbage, Inc. He was a co-founder of iSpheres in the area of event-driven architecture; that software is now marketed by Avaya. Dr. Chandy does research on sense-and-respond systems. He has published three books and more than 100 papers on distributed computing, verification of concurrent programs, parallel programming languages, and performance models of computing and communication systems. Dr. Chandy received his Ph.D. from the Massachusetts Institute of Technology in electrical engineering at the Operations Research Center in 1969. He received a master's from the Polytechnic Institute of Brooklyn, and a bachelor's from the Indian Institute of Technology, Madras, in 1965.

Dennis S. Mileti is a recently retired professor and former chair of the Department of Sociology at the University of Colorado, Boulder, and director emeritus of the Natural Hazards Center. Dr. Mileti is author of more than 100 publications, most of which focus on the societal aspects of mitigation, preparedness, response, and recovery for hazards and disasters. His book *Disasters by Design* (1999) involved more than 130 experts to assess knowledge, research, and policy needs for hazards in the United States. He has served on a variety of advisory boards and was co-founder and co-editor-in-chief of *Natural Hazards Review,* an interdisciplinary all-hazards journal devoted to bringing together the natural and social sciences, engineering, and the policy communities. Dr. Mileti received his Ph.D. in sociology from the University of Colorado, Boulder.

Helena Mitchell is the executive director of the Center for Advanced Communications Policy and principal research scientist at the Georgia Institute of Technology. In tandem, she is also the principal investigator (PI) for the Rehabilitation Engineering Research Center for Wireless Technologies, funded by the U.S. Department of Education since 2001 to promote equitable access to wireless technologies by people with disabilities and the adoption of universal design in wireless devices. Dr. Mitchell was recruited to Georgia through the Georgia Research Alliance Eminent Scholar program that spans educational, community, and business environments. Her areas of specialty include broadband and wireless communications, educational technologies, regulatory and legislative policy, emergency/public safety communications, and universal service to vulnerable, rural, and inner-city populations. Dr. Mitchell has held positions in academia, business, and government, which contribute to her unique ability to see multiple perspectives. This expertise has enabled her to create innovative interdisciplinary technology and educational programs, as well as utilize her unique skill for navigating new waters. Dr. Mitchell has held executive posts in Washington, D.C., with the federal government. At the FCC, she served as the associate chief, strategic communications, for the Office of Engineering and Technology to increase commission dialog with advanced technology companies. Earlier, as the chief of the Emergency Broadcast System (EBS), her work resulted in major rulemakings that expanded EBS to include cable, satellite, and advanced communications systems and the adoption of the Emergency Alert System. As a result, her team was selected as the FCC Organization of the Year. Dr. Mitchell previously headed the telecommunications development programs for the National Telecommunications and Information Administration of the U.S. Department of Commerce, where she spearheaded executive branch policy initiatives to increase educational, broadcast, and nonbroadcast telecommunications ownership opportunities; advanced joint venture

APPENDIX C 63

projects between the education and business sectors; worked on international privatization activities; and was responsible for earmarking more than $50 million dollars in domestic and international grants and loans. In recognition of the success of her policy initiatives in telecommunications, she received the prestigious U.S. Department of Commerce Silver Medal. Dr. Mitchell received her Ph.D. in telecommunications policy from Syracuse University.

Ramesh R. Rao is the director of the University of California, San Diego (UCSD), division of the California Institute for Telecommunications and Information Technology (Calit2). In 2004, he was appointed the first holder of the Qualcomm Endowed Chair in Telecommunications and Information Technologies in the Department of Electrical and Computer Engineering of the Jacobs School of Engineering at UCSD, where he has been a faculty member since 1984. Prior to becoming the Calit2 UCSD division director in 2001, he served as the director of UCSD's Center for Wireless Communications. In addition to directing Calit2, Dr. Rao is involved on a day-to-day basis with a wide variety of interdisciplinary and collaborative research initiatives, leading several major projects at Calit2. He has been a lead investigator on dozens of major federal-, state-, foundation-, defense-, and industry-funded grants, including the National Institutes of Health-funded Wireless Internet Information System for Medical Response in Disasters Self-Scaling Systems for Mass Casualty Management, the Multimedia Telemedical Diagnostic System, the National Science Foundation (NSF)-funded Responding to Crises and Unexpected Events and ResponSphere projects, and multiple projects involving cognitive networking, as well as leading several international collaborations. He has authored more than 230 peer-reviewed technical papers on a wide range of research topics in wireless communications, including architectures, protocols, performance analysis of computer and communication networks, adaptive systems, energy-efficient communications, disaster management applications, and health-related applications, among others. He is currently engaged in numerous projects to bridge emerging technologies with medicine and healthcare and investigating the power of utilizing information technologies to enhance, even transform, healthcare resources, knowledge bases, and outcomes. Dr. Rao received his Ph.D. in computer science from the University of Maryland, College Park.

Shashi Shekhar is a McKnight Distinguished University Professor at the University of Minnesota (computer science faculty). For contributions to geographic information systems (GIS), spatial databases, and spatial data mining, he received the IEEE-Computer Society (CS) Technical Achievement Award and was elected an IEEE fellow as well as an American Association for the Advancement of Science fellow. He was also named a key difference-maker for the field of GIS by the most popular GIS textbook. He has a distinguished academic record that includes more than 260 refereed papers, a popular textbook *Spatial Databases* (2003), and an authoritative *Encyclopedia of GIS* (2008). Dr. Shekhar is serving as a member of the Computing Community Consortium Council (2012-2015), a co-editor-in-chief of *Geo-Informatica: An International Journal on Advances in Computer Sciences for GIS*, a series editor for the Springer-Briefs on GIS, and as a program co-chair for the International Conference on Geographic Information Science (2012). Earlier, Dr. Shekhar served on multiple National Research Council committees, including Future Workforce for Geospatial Intelligence (2011), Mapping Sciences (2004-2009), and Priorities for GEOINT Research (2004-2005). He also served as a general co-chair for the Internaional Symposium on Spatial and Temporal Databases (2011) and the Association for Computing Machinery Geographic Information Systems (1996). He also served on the board of directors of University Consortium on GIS (2003-2004) and was on the editorial board of *IEEE Transactions on Knowledge and Data Engineering* and the IEEE-CS Computer Science and Engineering Practice Board. In early 1990s, Dr. Shekhar's research developed core technologies behind in-vehicle navigation devices as well as web-based routing services, which revolutionized outdoor navigation in urban environments in the last decade. His recent research results played a critical role in evacuation route planning for homeland security and received multiple recognitions including the Center for Transportation Studies Research Partnership Award for significant impact on transportation. He pioneered the research area of spatial data mining via pattern families (e.g., collocation, mixed-drove co-occurrence, cascade), keynote speeches, survey papers, and workshop organization. Dr. Shekhar received a Ph.D. degree in computer science from the University of California, Berkeley.

Ming-Hsiang (Ming) Tsou is a professor in the Department of Geography, San Diego State University. He received a B.S. from National Taiwan University in 1991, an M.A. from the State University of New York at Buffalo in 1996, and a Ph.D. from the University of Colorado, Boulder, in 2001, all in geography. His research interests are in mapping cyberspace and social media, Internet mapping, Web GIS applications, mobile GIS and wireless communication, and cyberinfrastructure with grid and cloud computing technology. He has applied his research interests in wildfire mapping, environmental monitoring and management, habitat conservation, K-12 education, and homeland border security. He is co-author of the book *Internet GIS* and has served on the editorial boards of *Annals of GIS* (since 2008) and *The Professional Geographer* (since 2011). Dr. Tsou was the chair of the Cartographic Specialty Group (2007-2008) and the chair of Cyberinfrastructure Specialty Group (2012-2013) in the Association of American Geographers. Dr. Tsou served on the 2006 committee on "Research Priorities for the USGS Center of Excellence for Geospatial Information Science." In 2007, he created and maintained an interactive Web-based mapping services for San Diego Wildfires 2007. In 2010, Dr. Tsou served as the PI of a NSF-CDI-funded project, "Mapping ideas from Cyberspace to Realspace."